Lecture Notes in Mathematics

Edited by A. Dold and B. Eckmann

149

Richard G. Swan

Notes by

E. Graham Evans

K-Theory of Finite Groups and Orders

Springer-Verlag
Berlin Heidelberg New York Tokyo

Author

Richard G. Swan
Department of Mathematics
University of Chicago, 5734 University Avenue
Chicago, Illinois 60637, USA

1st Edition 1970
2nd Corrected Printing 1986

ISBN 3-540-04938-X Springer-Verlag Berlin Heidelberg New York Tokyo
ISBN 0-387-04938-X Springer-Verlag New York Heidelberg Berlin Tokyo

Printing and binding: Beltz Offsetdruck, Hemsbach/Bergstr.
2146/3140-543210

NOTE

Throughout these notes, the lower case German "p" has been consistently typed as a lower case Roman "y". This results in some rather unusual notation , but should not cause any difficulty if the reader is prepared for it. The lower case German "c" is also indistinguishable from the lower case Greek tau, but again no real confusion should result.

TABLE OF CONTENTS

Chapter One: Introduction

Let R be a ring. Then $K_0(R)$ is the abelian group given by generators [P] where P is a finitely generated projective R module, with relations $[P] = [P'] + [P'']$ whenever $0 \to P' \to P \to P'' \to 0$ is an exact sequence of finitely generated projective R modules. K_0 is a covariant functor from rings to abelian groups. If $f: R \to R'$, then $K_0(f'): K_0(R) \to K_0(R')$ by $[P] \to [R' \otimes_R P]$. If R is left noetherian, then $G_0(R)$ is the abelian group with generators [M] where M is a finitely generated left R module with relations $[M] = [M'] + [M'']$ whenever $0 \to M' \to M \to M'' \to 0$ is an exact sequence of finitely generated left R modules. G_0 is not a functor since the tensor product $R' \otimes_R$ will preserve all the relations only when R' is flat as a right R module. There is a map $K_0(R) \to G_0(R)$ given by $[P] \rightsquigarrow [P]$. This map is called the Cartan map. It is natural with respect to maps of rings $R \to R'$ such that R' a flat right R module. If R is left artinian then $G_0(R)$ is free abelian on $[S_1], \ldots, [S_n]$ where S_i are the distinct classes of simple R modules. $K_0(R)$ is free abelian on $[I_1], \ldots, [I_r]$ where I_i are the distinct classes of indecomposable projective R modules.

The map $K_0(R) \to G_0(R)$ gives a matrix (a_{i_j}) where a_{i_j} = the number of times S_j occurs in a composition series for I_i.

Definition. A ring R is left regular if

1) R is left noetherian, and

2) every finitely generated left R module has a finite resolution by finitely generated projective left R modules.

If R is commutative, we omit the adjective "left".

We recall the following theorem whose proof is well known [SK] and is essentially the same as that of Theorem 1.2 below.

Theorem 1.1. If R is left regular, then the Cartan map is an isomorphism.

Examples (Bass-Murty). The Cartan map is neither a monomorphism nor an isomorphism in general. If π is a finitely generated abelian group, and $Z\pi$ is its integral group ring, then $G_0(Z\pi)$ is finitely generated but $K_0(Z\pi)$ is not finitely generated in general. If π is a finite group, then $K_0(Z\pi) = Z \oplus$ finite group but $G_0(Z\pi)$ has rank greater than 1 in general.

Definition. Let R be a commutative ring and A an R algebra. $G_0^R(A)$ is the abelian group with generators [M] where M is a left A module which is finitely generated and projective as an R module, with relations [M] = [M'] + [M"] for each A exact sequence $0 \to M' \to M \to M'' \to 0$ of A modules which are finitely generated as R modules.

Theorem 1.2. Let R be a regular commutative ring and A an R algebra which is finitely generated and projective as an R module. Then the map $f\colon G_0^R(A) \to G_0(A)$ given by $[M] \rightsquigarrow [M]$ is an isomorphism.

Remark. The examples of main concern are integral group rings of finite groups.

Proof. First we need a few preliminary results.

<u>Lemma 1.3.</u> (Schanuel) Let R be a ring and $0 \to B \to P \to A \to 0$ and $0 \to B' \to P' \to A \to 0$ be two exact sequences of R modules with P and P' projective. Then $P \oplus B'$ is isomorphic to $P' \oplus B$.

<u>Proof.</u> Consider the diagram

$$\begin{array}{ccccccccc}
 & & & & 0 & & 0 & & \\
 & & & & \downarrow & & \downarrow & & \\
 & & & & B & = & B & & \\
 & & & & \downarrow & & \downarrow & & \\
0 & \to & B' & \to & X & \to & P & \to & 0 \\
 & & \| & & \downarrow & & \downarrow & & \\
0 & \to & B' & \to & P' & \to & A & \to & 0 \\
 & & & & \downarrow & & \downarrow & & \\
 & & & & 0 & & 0 & &
\end{array}$$

where X is the pullback. Then since P and P' are projective X is isomorphic to $P \oplus B'$ and to $P' \oplus B$.

<u>Corollary 1.4.</u> Let R be a ring and

$0 \to B_n \to P_{n-1} \to \ldots \to P_0 \to A \to 0$ and

$0 \to B_n' \to P_{n-1}' \to \ldots \to P_0' \to A \to 0$ be exact sequences of R modules with P_i and P_i' all projective. Then

$B_n \oplus P_{n-1}' \oplus P_{n-2} \oplus \ldots.$ is isomorphic to $B_n' \oplus P_{n-1} \oplus P_{n-2}' \oplus \ldots$.

<u>Proof.</u> This follows by an easy induction argument from lemma 1.3.

<u>Remark</u> (I. Kaplansky). Corollary 1.4 enables one to define the homological dimension of a module (without reference to Ext) as the first n such that the kernel of any n step projective resolution is projective. By Corollary 1.4, we get the same n for any projective resolution.

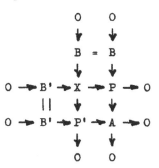

We return to theorem 1.2. We want to define an inverse, g, for f: $G_0^R(A) \to G_0(A)$. It clearly is enough to define g on generators. Let M be a finitely generated A module. Then M is a finitely generated R module and has finite homological dimension over R. Say $hd_R M = n$. Let $0 \to B_n \to P_{n-1} \to \dots \to P_j \to M \to 0$ be an A exact sequence with all the P_i finitely generated projective A modules. Then all the P_i are finitely generated projective R modules since A is a finitely generated projective R module. Hence B_n is a finitely generated projective R module. We define $g([M]) = [P_0] - [P_1] + \dots (-1)^n[B_n]$. We must show that g is well defined on generators and preserves relations. Let

1) $\quad 0 \to B'_m \to P'_{m-1} \to \dots \to P'_0 \to M \to 0$

be an A exact sequence with P'_i finitely generated projective A modules with $m \geq n$. Let

2) $\quad 0 \to B_m \to P_{m-1} \to \dots \to P_n \to B_n \to 0$

be an exact sequence with all P finitely generated projective A modules. Then

3) $\quad 0 \to B_n \to P_{m-1} \to \dots \to P_n \to P_{n-1} \to \dots \to P_0 \to M \to 0$

is an exact sequence. Apply corollary 1.4 to 1) and 3). Then we want to show that 2) implies that
$[B_m] - [P_{m-1}] + \dots \pm [P_n] \mp [B_n] = 0$ in $G_0^R(A)$. That is we must

show that if $0 \to M_p \to \ldots \to M_0 \to 0$ is an A exact sequence of finitely generated projective R modules then $\sum_i (-1)^i [M_i] = 0$.

We prove this by induction on n by making two A exact sequences $0 \to M_k \to \ldots \to M_2 \to X \to 0$ and

4) $\quad 0 \to X \to M_1 \to M_0 \to 0$

and noting that 4) implies that the A module X is a finitely generated projective R module since 4) splits as a sequence of R modules.

Now given two resolutions of M, we can assume they have the same length by the result just proved. Corollary 1.4 now shows that g is well defined on generators.

Let $0 \to M' \to M \to M'' \to 0$ be an A exact sequence of finitely generated A modules. Pick $n \geq hd_R M', hd_R M, hd_R M''$. Pick n step resolutions of M' and M" by finitely generated projective A modules. Combine to get an n step resolution of M (see Cartan-Eilenberg, p. 80) giving the following commutative diagram of A modules with all rows and columns exact:

$$\begin{array}{ccccccccc}
 & & 0 & & 0 & & 0 & & \\
 & & \downarrow & & \downarrow & & \downarrow & & \\
0 & \to & P'_n & \to & \ker h & \to & P''_n & \to & 0 \\
 & & \downarrow & & \downarrow h & & \downarrow & & \\
 & & \vdots & & \vdots & & \vdots & & \\
0 & \to & P'_0 & \to & P'_0 \oplus P''_0 & \to & P''_0 & \to & 0 \\
 & & \downarrow & & \downarrow & & \downarrow & & \\
0 & \to & M' & \to & M & \to & M' & \to & 0 \\
 & & \downarrow & & \downarrow & & \downarrow & & \\
 & & 0 & & 0 & & 0 & &
\end{array}$$

where all P_i', P_i'' are finitely generated projective A modules. Then $g([M]) = g([M']) + g([M''])$ since $g([N])$ can be defined from any projective resolution of length greater than or equal to $hd_R N$ where N is a finitely generated A module.

Hence g gives a well defined map $g\colon G_0(A) \to G_0^R(A)$. g is clearly the inverse of f.

Using theorem 1.2 we will investigate the functorality of G_0 and G_0^R. Let R be a regular commutative ring and $h\colon R \to R'$ a homomorphism of commutative rings. Let π be any finite group. Then we have a homomorphism h^* of group rings induced by h, $h^*\colon R\pi \to R'\pi$. $R' \otimes_R$ __ induces a map $K_0(R\pi) \to K_0(R'\pi)$.

We also have the diagram

$$\begin{array}{ccc} G_0(R\pi) & & G_0(R'\pi) \\ \uparrow f & & \uparrow f' \\ G_0^R(R\pi) & \xrightarrow{\ k\ } & G_0^{R'}(R'\pi) \end{array}$$

where f is an isomorphism and k is defined by $k([M]) = [R' \otimes_R M]$ which preserves relations since we are only considering modules which are projective over R, and, hence, all short exact sequences split over R. If g is the inverse of f, then $f'kg\colon G_0(R\pi) \to G_0(R'\pi)$. Therefore, $G_0(R\pi)$ is a functor on the category of regular commutative rings. The main use of this will be for the cannonical projection $Z \to Z/pZ$.

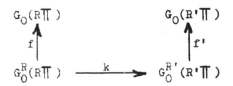

If M and N are finitely generated $R\pi$ modules, then $M \otimes_R N$ is a finitely generated $R\pi$ module where $p(m \otimes n) = (pm \otimes pn)$ for all $p \in \pi$. If M', M, and M" are $R\pi$ modules which are finitely generated and projective as R modules, if N is any $R\pi$ module, and if the $R\pi$ sequence $0 \to M' \to M \to M'' \to 0$ is exact then the $R\pi$ sequence $0 \to M' \otimes_R N \to M \otimes_R N \to M'' \otimes_R N \to 0$ is exact. If N is also a finitely generated projective R module then so are $M' \otimes_R N$, $M'' \otimes_R N$, and $M \otimes_R N$. Hence there is a well defined map

$$G_0^R(R\pi) \times G_0^R(R\pi) \longrightarrow G_0^R(R\pi) \text{ given}$$

by $[M] \times [N] \rightsquigarrow [M \otimes_R N]$. If we fix N then we have $h_N\colon G_0^R(R\pi) \to G_0^R(R\pi)$ given by $[M] \rightsquigarrow [M \otimes_R N]$. If $0 \to N' \to N \to N'' \to 0$ is exact then $h_N = h_{N''} + h_{N'}$. Since the π action in $M \otimes_R N$ is the diagonal this product is associative and commutative. The $R\pi$ module R with trivial π action is clearly the identity. Hence we have proved:

Theorem 1.5. Let R be a commutative ring and π a finite group. Then $G_0^R(R\pi)$ is an associative commutative ring with identity.

Corollary 1.6. If in the above R is regular, then $G_0(R\pi)$ is an associative commutative ring with 1.

By the same proof, we can show

Corollary 1.7. With the notation of theorem 1.5 $K_0(R\pi)$ is an associative commutative ring without 1 (unless $\pi = \{1\}$ since otherwise R is not projective over $R\pi$).

If $R \to R'$ is a homomorphism of rings such that R' is a finitely generated R module then $[M] \rightsquigarrow [M]$ gives a well defined map $G_0(R') \to G_0(R)$. Let R be a commutative noetherian ring, A an R algebra which is finitely generated as an R module, and S a multiplicatively closed set of R containing 1 but not containing 0. Then A_S is an algebra over R_S and $M \rightsquigarrow R_S \otimes_R M_S = M$, induces a well defined map $G_0(A) \to G_0(A_S)$. If M is a finitely generated A module such that $M_S = 0$ then there exists an $s \in S$ such that $sM = 0$. Hence M can be thought of as an $A/(s)$ module. Hence the map $G_0(A) \to G_0(A_S)$ annihilates the images of all $G_0(A/(s))$ for $s \in S$. The following two theorems describe in full detail the map $G_0(A) \to G_0(A_S)$. The proofs may be found in [SK, Chapter 5].

Theorem 1.6. Let R be a commutative noetherian ring, A an R algebra finitely generated as an R module, S a multiplicatively closed set containing 1 but not 0. Then the sequence

$\coprod_{s \in S} G_0(A/(s)) \to G_0(A) \to G_0(A_S) \to 0$ is exact.

Theorem 1.7. With the same notation as the above the sequence $\coprod_{i \in I} G_0(A/p_iA) \to G_0(A) \to G_0(A_S) \to 0$ is exact where $\{p_i\}$ is the set of all prime ideals of R with $p_i \cap S \neq \emptyset$.

Let k be a field and A a k algebra which is finitely generated as a k module then $G_0(A)$ is free abelian on $[S_1], \ldots, [S_n]$ where the S_i represent all the isomorphism classes of simple A modules. If M is a finitely generated A module, then

$[M] = \sum_{i=1}^{n} a_i[S_i]$ $a_i \in Z$ in $G_0(A)$. We want to recover the a_i from M.

We could try characters, but they only yield information up to the characteristic of k. Pick $a \in A$. Then $\hat{a}: M \to M$ is given by $\hat{a}(m) = am$ for all $m \in M$. We define $\varphi_a(M)$ = characteristic polynomial of $\hat{a}$ on M. Then $\varphi_a(M) \in k[X]$. If $0 \to M' \to M \to M'' \to 0$ is an exact sequence of finitely generated A modules, then $\varphi_a(M) = \varphi_a(M')\,\varphi_a(M'')$. For, pick a basis of $M' \subset M$ and extend to a basis for M, which projects to a basis for M". Computing the determinant involved gives the result. If $M \neq 0$ then $\varphi_a(M) \neq 0$. For, if $\dim_k M = n$ $\varphi_a(M) = X^n + \dots$. If $M = 0$, define $\varphi_a(M) = 1$. Now $k[x] - \{0\} = k[x]^+$ is contained in $k(x) - \{0\} = k(x)^*$. Define $\varphi_a\colon G_0(A) \to k(X)^*$ to be the homomorphism sending addition in $G_0(A)$ to multiplication in $k(X)^*$ such that $[M] \mapsto \varphi_a(M)$.

<u>Theorem 1.8</u> (Brauer). If $x \in G_0(A)$ and $\varphi_a(X) = 1$ for all $a \in A$, then $x = 0$.

<u>Remark</u>. We can even pick out a finite number of $a \in A$ such that it is enough to check for them.

<u>Proof</u>. $x = \sum k_i[S_i] = \sum_{m_i>0} m_i[S_i] - \sum_{n_i>0} n_i[S_i]$ where each isomorphism class $[S_i]$ occurs only once in the sum. Let $M = \coprod_{m_i>0} S_i^{m_i}$ and $N = \coprod_{n_i>0} S_i^{n_i}$. Then $x = [M] - [N]$ where M and N are semisimple A modules. $\varphi_a(X) = 1 = \varphi_a(M)\varphi_a(N)^{-1}$ implies that $\varphi_a(M) = \varphi_a(N)$. Hence, it is enough to show that a semisimple A module is determined up to isomorphism by the values $\varphi_a(M)$ $a \in A$.

Let R be the radical of A. Then RM = 0 since M is semisimple. Thus, M is an $A/R = \bar{A}$ module and $\bar{A}$ is semisimple since A has finite dimension over k. Hence $\bar{A} = B_1 \times \ldots \times B_n$ where the B_i are simple algebras. A acts on S_i through B_i. Let e_i be the identity of B_i. Then the e_i are primative central idempotents and e_i acts on S_j as $\{\begin{smallmatrix} 1 \text{ if } i = j \\ 0 \text{ if } i \neq j \end{smallmatrix}$. Lift the e_i to a_i in A. We claim the $\varphi_{a_i}(M)$ determine M. a_i acts on S_j in the same way as e_i. Say $M = \coprod_{j \neq i} \_\_\_\_ \oplus \coprod_1^{m_i} S_i$. Then a_i acts like 0 on the first summand and like 1 in the second. Hence $\varphi_{a_i}(M) = X^r\,(X-1)^{m_i(\dim_k S_i)}$. Thus we can recover the m_i from $\varphi_{a_i}(M)$ and M from the collection of the $\varphi_{a_i}(M)$.

<u>Theorem 1.9</u>. Let R be a regular commutative domain, A an R algebra which is a finitely generated projective R module, m a maximal ideal of R, and K the quotient field of R. There is a unique map $f\colon G_0(K \otimes_R A) \to G_0(A/mA)$ which makes the following diagram commute:

$$\begin{array}{ccccc} G_0^R(A) & \longrightarrow & G_0(A) & \longrightarrow & G_0(K \otimes_R A) \\ & & & \searrow & \vdots\, f \\ & & & & G_0(A/mA) \end{array}$$

Remarks. The hypothesis insures that all maps exist since $G_0^R(A) \to G_0(A)$ is an isomorphism. f is given by a matrix which is called Brauer's matrix of decomposition. By theorem 1.6 or 1.7 the map $G_0(A) \to G_0(K \otimes_R A)$ is onto and hence f, if it exists, is unique.

Proof. Let $S = R - \{0\}$. Then $K = R_S$ and $K \otimes_R A = A_S$. Let M be a finitely generated A module such that $sM = 0$ for some $s \in S$. If $[M] \rightsquigarrow 0$ in $G_0(A/mA)$ for all such M, then f exists by theorem 1.6 or 1.7. Let 1) $0 \to P_n \to \ldots \to P_0 \to M \to 0$ be an A exact sequence with all the P_i finitely generated and projective over R. Then $[M] \in G_0(A)$ corresponds to $\sum (-1)^i [P_i] \in G_0^R(A)$. Since $A/mA = A_m/m_mA_m$ we can factor the map $G_0(A) \to G_0(A/mA)$ as $G_0(A) \to G_0(A_m) \to G_0(A/mA)$. That is we can assume R is local. Thus all the P_i are free over R.

Localizing the sequence 1) at S we get $0 \to P_{hS} \to \ldots \to P_{0S} \to M_S \to 0$ and $M_S = 0$. Therefore, $\sum (-1)^i [P_i]_S = 0$ in $G_0(A_S)$. Hence for each $a \in A$, $\prod \varphi_a(P_{iS})^{(-1)^i} = 1$. Since P_i are all free over R, we pick a basis for each and obtain matrices for $\hat{a}$ on the P_i which are the same matrices as for $\hat{a}$ on the P_{iS}. Therefore, $F_i = \varphi_a(P_{iS})$ = characteristic polynomial $P_i \xrightarrow{\hat{a}} P_i$ is in R[X]. Hence by theorem 1.8 $\prod_{i \text{ even}} \varphi_a(P_i) = \prod_{i \text{ odd}} \varphi_a(P_i)$. Therefore $\prod_{i \text{ even}} F_i = \prod_{i \text{ odd}} F_i$. Reducing mod m we have $\sum (-1)^i [P_i/mP_i] \in G_0(A/mA)$. The bases for P_i reduced mod m give bases for P_i/mP_i since R is local with maximal ideal m. Let $a \in A$ and reduce mod m to get $\bar{a} \in \bar{A} = A/mA$. Then the matrix for the characteristic polynomial for $\hat{\bar{a}}$, $\varphi_{\bar{a}}(P_i/mP_i)$, is $\varphi_a(P_i/mP_i)$ reduced mod m. But reducing mod m is a ring homomorphism. Hence $\prod_{i \text{ odd}} \varphi_{\bar{a}}(P_i/mP_i) = \prod_{i \text{ even}} \varphi_{\bar{a}}(P_i/mP_i)$ since

$\prod_{i\ \mathrm{odd}} F_i/mF_i = \prod_{i\ \mathrm{even}} F_i/mF_i$. Thus $[M] \rightsquigarrow \sum (-1)^i[P_i/mP_i] = 0$ in $G_0(A/mA)$. Hence the map f exists.

Theorem 1.10. Let R be a commutative local domain (not necessarily noetherian) with quotient field K and maximal ideal m. Let A be an R algebra finitely generated and projective as an R module. Assume that the Cartan map κ: $K_0(A/mA) \rightarrow G_0(A/mA)$ is a monomorphism. Let P and Q be finitely generated projective A modules such that $K \otimes_R P$ is isomorphic to $K \otimes_R Q$ over $K \otimes_R A$. Then P is isomorphic to Q.

Remarks. We will show later that the Cartan map is a monomorphism for $R\Pi$ where Π is a finite group. This theorem is false even for ideals of Dedekind rings if R is not local.

Proof. P and Q are finitely generated free R modules. Pick bases $p_1, \dots, p_n$ and $q_1, \dots, q_n$ where $n = \dim_K K \otimes_R P = \dim_K K \otimes_R Q$. For each $a \in A$ we have $\hat{a}$: $P \rightarrow P$ and $\hat{a}$: $Q \rightarrow Q$ yielding matrices A and B with respect to the bases $\{p_i\}$ and $\{q_i\}$. Let f and g be the characteristic polynomials of A and B. Since $K \otimes_R P$ and $K \otimes_R Q$ are isomorphic $f = g$. $\bar{P} = P/mP$ is a module over A/mA. As a $k = R/m$ module $\bar{P}$ is free on $\{\bar{p}_i\}$, the images of $\{p_i\}$. As before the characteristic polynomial of $\hat{\bar{a}}$ on P is f with coefficients reduced mod m. Hence for every $\bar{a} \in A/mA$ the characteristic polynomials for $\bar{P}$ and $\bar{Q}$ are the same. Thus, by theorem 1.8 $[\bar{P}] - [\bar{Q}] = 0$ in $G_0(A/mA)$. That is if $X \in G_0^R(A)$ goes to 0 in $G_0(K \otimes_R A)$ then X goes to 0 in $G_0(A/mA)$. Thus $\kappa([\bar{P}] - [\bar{Q}]) = 0$ in $G_0(A/mA)$. But κ is a monomorphism. Hence $[\bar{P}] - [Q] = 0$ in $K_0(A/mA)$. Hence $P/mP \oplus F$

is isomorphic to $Q/mQ \oplus F$ where F is a finitely generated free A/mA module. A/mA is artinian. Therefore, the Krull-Schmidt theorem applies and we cancel the F's getting P/mP isomorphic to Q/mQ.

Thus we have shown $K \otimes_R P$ isomorphic to $K \otimes_R Q$ implies P/mP is isomorphic to Q/mQ. We examine the diagram

$$\begin{array}{ccc} P & \overset{f}{\cdots\cdots\to} & Q \\ \downarrow & & \downarrow \\ P/mP & \xrightarrow{f'} & Q/mQ \\ \downarrow & & \downarrow \\ 0 & & 0 \end{array}$$

where f' is any isomorphism of $\bar{P}$ with $\bar{Q}$. f exists since P is projective and maps onto Q/mQ. f is onto by Nakayama's lemma. Q is projective and hence f splits. Now $N = \ker f$ is finitely generated since it is a direct summand of Q. Since f splits, $N/mN = \ker f' = 0$, and hence N is 0 by Nakayama's lemma. Hence f is an isomorphism and P and Q are isomorphic. More quickly, we note that P and Q are projective covers for $\bar{P}$ and $\bar{Q}$ which are isomorphic and hence P and Q are isomorphic.

Chapter Two: Frobenius Functors

In this chapter we discuss $G_0(R\Pi)$ and $K_0(R\Pi)$ as functors in Π.

Let Π be a finite group, Π' a subgroup, and $i: \Pi' \to \Pi$ the inclusion. Let $\Pi = \bigcup \Pi' g_i \quad g_i \in \Pi$ be the decomposition of Π

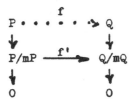

into disjoint cosets. Then $R\pi$ is a free $R\pi'$ module with base $g_1, \ldots, g_n$. We have $G_0(R\pi') \xleftarrow{i^*} G_0(R\pi)$ given by $i^*([M]) = [M]$. This is functorial for category of finite groups and inclusion maps. Since $R\pi$ is a free $R\pi'$ module, it is flat and hence we have $G_0(R\pi') \xrightarrow{i_*} G_0(R\pi)$ given by $i_*[M] = [R\pi \otimes_{R\pi'} M]$. This was defined by Frobenius long before tensor products were known. It was called induction or forming the induced representation.

We assume R is a regular commutative ring. Then $G_0(R\pi) = G_0^R(R\pi)$ are all rings. For any map of finite groups $i\colon \pi' \to \pi$ i^* is defined and is a ring homomorphism where we use the ring structure on $G_0^R(R\pi')$ and push it down to $G_0(R\pi')$ via the isomorphism. Even without regularity $i^*\colon G_0^R(R\pi) \to G_0^R(R\pi')$ is a ring homomorphism. However $i_*\colon G_0^R(R\pi') \to G_0^R(R\pi)$ is not a ring homomorphism. For $i_*([M]) = [R\pi \otimes_{R\pi'} M]$ and $R\pi \otimes_{R\pi'} M$ is isomorphic to $\coprod_{1}^{[\pi:i(\pi')]} M$ as an R module. Hence i_* does not preserve the identity as a rank count shows. But i_* does give a covariant functor from finite groups and monomorphisms to abelian groups.

<u>Theorem 2.1</u>. (Frobenius) Let R be a commutative ring, $i\colon \pi' \to \pi$ an inclusion of finite groups, $x \in G_0^R(R\pi)$, and $y \in G_0^R(R\pi')$. Then $i_*(i^*(x)y) = xi_*(y)$.

<u>Remarks</u>. This says $i_*\colon G_0^R(R\pi') \to G_0^R(R\pi)$ is a homomorphism of $G_0^R(R\pi)$ modules where $G_0^R(R\pi')$ is a $G_0^R(R\pi)$ module through the action

of i^*. The formula is similar to the relation between homology and cohomology.

Proof. We prove this by proving the following:

Theorem 2.2. With the same notation as above let M be an $R\pi$ module and N an $R\pi'$ module. Then $R\pi \otimes_{R\pi'} (M \otimes_R N)$ is isomorphic to $M \otimes_R (R\pi \otimes_{R\pi'} N)$ where M is considered as an $R\pi'$ module on the left side.

Proof. We define a map f on generators which will be an isomorphism. Let $p \otimes (m \otimes n) \in R\pi \otimes_{R\pi'} (M \otimes_R N)$. We define $f(p \otimes (m \otimes n)) = pm \otimes (p \otimes n)$. f is automatically an R map. Let $q \in \pi$. Then $q(f(p \otimes (m \otimes n)) = qpm \otimes (qp \otimes n) = f(q(p \otimes (m \otimes n)))$. Hence f is an $R\pi$ map. Let $q \in \pi'$, then $pi(q) \otimes m \otimes n = p \otimes qm \otimes n$ and $f(pi(q) \otimes m \otimes n) = pi(q)m \otimes pi(q) \otimes n = p(qm) \otimes (p \otimes qn) = f(p \otimes qm \otimes n)$. Hence f is well defined over action of $R\pi'$ in $R\pi \otimes_{R\pi'} (M \otimes_R N)$. We define an inverse g to f by $g(m \otimes (p \otimes n)) = p \otimes (p^{-1}m \otimes n)$. This is clearly multilinear over R we need it is over π'. $g(m \otimes (pi(q) \otimes n)) = pi(q) \otimes (i(q)^{-1}p^{-1}m \otimes n) = p \otimes (p^{-1}m \otimes qn) = g(m \otimes (p \otimes qn))$ for all $q \in \pi'$. g and f are clearly inverses. This completes theorems 2.1 and 2.2.

Definition (Lam). A Frobenius functor is

1) a contravariant functor F: finite groups and monomorphisms $\longrightarrow$ commutative rings with 1 and homomorphisms preserving 1.

2) For each $i: \pi' \to \pi$ an inclusion of finite groups we are given $i_*: F(\pi') \to F(\pi)$ such that i_* makes F into a covariant functor from finite groups and monomorphisms to abelian groups. i_* is additive, $(ij)_* = i_*j_*$, and $1_* = 1$.

3) $i: \pi' \to \pi$. Let $x \in F(\pi)$ and $y \in F(\pi')$. Then $i_*(i^*(x)y) = xi_*(y)$.

<u>Definition</u>. A <u>morphism</u> α between two Frobenius functors F and G is defined by giving a ring homomorphism $\alpha_\pi : F(\pi) \to G(\pi)$ for each finite group π such that if $i: \pi' \to \pi$ is a monomorphism of finite groups then the diagrams

$$\begin{array}{ccc} F(\pi') & \xrightarrow{i_*} & F(\pi) \\ {\scriptstyle \alpha_{\pi'}}\downarrow & & \downarrow{\scriptstyle \alpha_\pi} \\ G(\pi') & \xrightarrow[i_*]{} & G(\pi) \end{array} \qquad \text{and} \qquad \begin{array}{ccc} F(\pi') & \xleftarrow{i_+} & F(\pi) \\ {\scriptstyle \alpha_{\pi'}}\downarrow & & \downarrow{\scriptstyle \alpha_\pi} \\ G(\pi') & \xleftarrow[i^*]{} & G(\pi) \end{array}$$

commute.

For example for the Frobenius functors $G_0^R(R\pi)$ and $G_0^{R'}(R'\pi)$, if $R \to R'$ is a homomorphism of commutative rings then $\alpha_\pi : G_0^R(R\pi) \to G_0^{R'}(R'\pi)$ given by $\alpha_\pi([M]) = [R' \otimes_R M]$ is a morphism.

If R is not regular, then $G_0(R\pi)$ fails to be a Frobenius functor since it is not a ring. If $i: \pi' \to \pi$ is a monomorphism of finite groups, then we have $G_0(R\pi) \underset{i_*}{\overset{i^*}{\rightleftarrows}} G_0(R\pi')$. Note that $K_0(R\pi)$ is a ring but without a unit. If $[P] \in K_0(R\pi)$ then

$[P] \in K_0(R\pi')$ since $R\pi$ is a finitely generated projective module over $R\pi'$. Hence we have i^* and i_* in this case also. Both $G_0(R\pi)$ and $K_0(R\pi)$ are modules over $G_0^R(R\pi)$. For if M is a finitely generated $R\pi$ module projective over R and N is a finitely generated $R\pi$ module, then $M \otimes_R N$ is a finitely generated $R\pi$ module. If $0 \to M' \to M \to M'' \to 0$ is an exact sequence of finitely generated $R\pi$ modules projective over R, then $0 \to M' \otimes_R N \to M \otimes_R N \to M'' \otimes_R N \to 0$ is exact since the first sequence splits over R. If $0 \to N' \to N \to N'' \to 0$ is an exact sequence of finitely generated $R\pi$ modules, then $0 \to M \otimes_R N' \to M \otimes_R N \to M \otimes_R N'' \to 0$ is exact since M is a projective (and hence flat) R module. Thus $[M] \times [N] = [M \otimes_R N]$ gives a well defined product making $G_0(R\pi)$ into a $G_0^R(R\pi)$ module. This suggests the following

<u>Definition</u> (Lam). Let F be a Frobenius functor. Then a <u>Frobenius module</u> M over F is defined by giving, for each finite group π, an $F(\pi)$ module $M(\pi)$ and, for each inclusion of groups $i: \pi' \to \pi$, two maps $i^*: M(\pi) \to M(\pi')$ and $i_*: M(\pi') \to M(\pi)$ such that

1) i_* and i^* are functorial and additive, i.e., M with i_* is an additive covariant functor and M with i^* is an additive contravariant functor.

2) If $x \in F(\pi)$ and $y \in M(\pi)$, then $i^*(xy) = i^*(x)i^*(y)$, and

3) if $x \in F(\pi)$ and $y \in M(\pi')$, then $i_*(i^*(x)y) = xi_*(y)$, and if $x \in F(\pi')$, $y \in M(\pi)$, then $i_*(xi^*(y)) = i_*(x)y$.

Examples. $G_0(R\Pi)$ is a Frobenius module over $G_0^R(R\Pi)$. If $F(\Pi)$ is a Frobenius functor, then $F(\Pi)$ is a Frobenius module over $F(\Pi)$.

Definition. If M and N are Frobenius modules over the Frobenius functor F then a morphism of Frobenius modules is defined by giving, for each Π, a morphism $f_\Pi : M(\Pi) \to N(\Pi)$ of $F(\Pi)$ modules which is natural with respect to i^* and i_*.

Frobenius modules and morphisms over a Frobenius functor F form an abelian category as the reader can easily check.

A more important example of a Frobenius module is $K_0(R\Pi)$ over $G_0^R(R\Pi)$. i_* and i^* are defined as usual. We need to check that if M is a finitely generated $R\Pi$ module projective over R and P a finitely generated projective $R\Pi$ module, then $M \otimes_R P$ is a finitely generated projective $R\Pi$ module. Then we will have a well defined multiplication as before. Since M and P are finitely generated over R, $M \otimes_R P$ will be finitely generated over R (and hence over $R\Pi$). Hence we only need the following:

Proposition 2.3. If M is an $R\Pi$ module which is projective (resp. free) as an R module and P is a projective (free) $R\Pi$ module, then $M \otimes_R P$ is a projective (free) $R\Pi$ module.

Proof. If M is free as an R module with base m_i and P is free as an $R\Pi$ module with base p_j, it is easy to see that $M \otimes_R P$ is free as an $R\Pi$ module with base $m_i \otimes p_j$. In the general case, we can find an $R\Pi$ module Q such that $P \oplus Q = F$ is a free $R\Pi$ module and an R module N such that $M \oplus N = G$ is a free R module. We let Π act

trivially on N making N an $R\Pi$ module. Then
$G \otimes_R F = (M \otimes_R P) \oplus (M \otimes_R Q) \oplus (N \otimes_R P) \oplus (N \otimes_R Q)$. Hence $M \otimes_R P$ is projective since $G \otimes_R F$ is free.

Thus $K_0(R\Pi)$ is a Frobenius functor. $K_0(R\Pi) \to G_0^R(R\Pi)$ given by $[P] \rightsquigarrow [P]$ and $G_0^R(R\Pi) \to G_0(R\Pi)$ given by $[M] \rightsquigarrow [M]$ are both morphisms of Frobenius modules. Hence the Cartan map $K_0(R\Pi) \to G_0(R\Pi)$ is also, so the kernel of the Cartan map is a Frobenius functor.

There are two basic types of morphisms of the Frobenius modules we have developed so far. The first is tensoring, the second forgetting. We cover them both in the next proposition.

<u>Proposition 2.4</u>. Let $R \to R'$ be a homomorphism of commutative rings. Then

1) $[M] \rightsquigarrow [R'\Pi \otimes_{R\Pi} M]$ induces a morphism of Frobenius modules over $G_0^R(R\Pi)$ in the cases

a) $G_0^R(R\Pi) \to G_0^{R'}(R'\Pi)$,

b) $G_0(R\Pi) \to G_0(R'\Pi)$ when R' is a flat R module, and

c) $K_0(R\Pi) \to K_0(R'\Pi)$.

2) $[M] \rightsquigarrow [M]$ induces a morphism of Frobenius modules over $G_0^R(R\Pi)$ in the cases

a) $G_0^R(R\Pi) \leftarrow G_0^{R'}(R'\Pi)$ when R' is a finitely generated projective R module,

b) $G_0(R\Pi) \leftarrow G_0(R'\Pi)$ when R' is a finitely generated projective R module, and

c) $K_0(R\Pi) \leftarrow K_0(R'\Pi)$.

<u>Proof</u>. This is routine for the most part. We only list a few facts which are necessary. We need to show that the maps are morphisms of modules for each R' and R and that the maps are natural with respect to both i^* and i_* induced from the inclusion $\Pi' \to \Pi$.

1) For example, we need that

$$\begin{array}{ccc} [M] & \longrightarrow & [R'\Pi \otimes_{R\Pi} M] \\ \downarrow & & \downarrow \\ [M \otimes_R N] & \longrightarrow & \end{array} \qquad \text{commutes.}$$

But this is true since $R'\Pi \otimes_{R\Pi} M$ is isomorphic to $R' \otimes_R M$, and, therefore, $(R'\Pi \otimes_{R\Pi} M) \otimes_{R'} N$ is isomorphic to $M \otimes_R N$.

2) Here we need that

$$\begin{array}{ccc} [N] & \longleftarrow & [N] \\ {\scriptstyle M \otimes_R -}\downarrow & & \downarrow{\scriptstyle R' \otimes_R M \otimes_{R'} -} \\ [M \otimes_R N] & \longleftarrow & [R' \otimes_R M \otimes_{R'} N] \end{array} \qquad \text{commutes.}$$

But it does because $R' \otimes_R M \otimes_{R'} N$ is isomorphic to $M \otimes_R N$. Thus, all maps in 1) and 2) are defined with respect to $G_0^R(R\Pi)$.

Next we check naturality. Let $\Pi' \to \Pi$ be an inclusion and let $*$ be G_0^R, G_0, or K_0.

First for i^* the maps

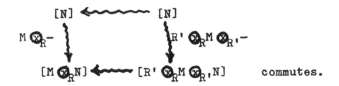

$$\begin{array}{ccc} \ast(R\Pi) & \longrightarrow & \ast(R'\Pi) \\ {\scriptstyle i^*}\downarrow & & \downarrow{\scriptstyle i^*} \\ \ast(R\Pi') & \longrightarrow & \ast(R'\Pi') \end{array} \quad \text{sends} \quad \begin{array}{ccc} M & \rightsquigarrow & R'\otimes_R M \\ \wr\downarrow & & \wr\downarrow \\ M & \rightsquigarrow & R'\otimes_R M \end{array}$$

which commutes.

For i_* we note that the maps

$$\begin{array}{ccc} \ast(R\Pi) & \longleftarrow & \ast(R'\Pi) \\ {\scriptstyle i_*}\uparrow & & \uparrow{\scriptstyle i_*} \\ \ast(R\Pi') & \longleftarrow & \ast(R'\Pi') \end{array} \quad \text{sends} \quad \begin{array}{ccc} R\Pi\otimes_{R\Pi'}M & \leftsquigarrow & R'\Pi\otimes_{R'\Pi'}M \\ \wr\uparrow & & \wr\uparrow \\ M & \longleftarrow & M \end{array}$$

So we need that $R'\Pi\otimes_{R'\Pi'}M$ is isomorphic to $R\Pi\otimes_{R\Pi'}M$ as a $R\Pi$ module. We could prove this by checking generators and relations or via an explicit map using the fact that $R'\Pi\otimes_{R'}M$ is isomorphic to $\coprod_{\sigma\in\Pi} M$ which is isomorphic to $R\Pi\otimes_R M$. But this was Frobenius' original construction for the tensor product. <u>Done</u>.

The reason for this work is to aid calculation of Frobenius functors. Suppose we are given a finite group Π and a Frobenius functor F. In general $F(\Pi)$ is hard to calculate. We may know it for Π cyclic however. So we want a method for obtaining information on $F(\Pi)$ from $\{F(\Pi')\}$ where Π' runs over the cyclic subgroups of Π.

More generally, let $\mathcal{C}$ be a class of groups, F a Frobenius functor, and M a module.

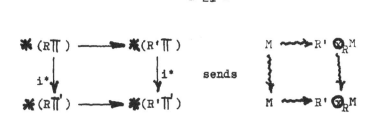

<u>Definitions</u>. $$F(\Pi)_{\mathcal{C}} = \sum_{\substack{\Pi' \in \mathcal{C} \\ \Pi' \subset \Pi}} i_* F(\Pi').$$

$$M(\Pi)_{\mathcal{C}} = \sum_{\substack{\Pi' \in \mathcal{C} \\ \Pi' \subset \Pi}} i_* M(\Pi').$$

If B is a group and A a subgroup of B then we say <u>A has exponent dividing n in B</u> (B/A has exp|n) if $x \in B$ implies $nx \in A$.

A typical theorem (to be proved later) is the following due to E. Artin:

<u>Theorem 2.5</u>. Let Π be a finite group of order n, K a field of characteristic 0 (so $K_0 = G_0 = G_0^R$), and $\mathcal{C}$ be the class of cyclic subgroups of Π, then $G_0(K\Pi)/G_0(K\Pi)_{\mathcal{C}}$ has exp|n.

We will use the machinery of Frobenius functors to extend Artin's Theorem.

<u>Lemma 2.6</u>. Let M be a Frobenius module over the Frobenius functor F then

a) $M(\Pi)_{\mathcal{C}}$ is a sub $F(\Pi)$ module of $M(\Pi)$,

b) $F(\Pi)_{\mathcal{C}}$ is an ideal of $F(\Pi)$, and

c) $F(\Pi)/F(\Pi)_{\mathcal{C}}$ is a ring and $M(\Pi)/M(\Pi)_{\mathcal{C}}$ is a module over it.

<u>Proof</u>.

a) $M(\Pi)_{\mathcal{C}}$ is closed under addition by definition and under multiplication by elements of $F(\Pi)$ by the definition of Frobenius module.

a) implies b) by taking M to be F.

c) We only need to check that $F(\Pi)_{\mathcal{C}}$ annihilates $M(\Pi)/M(\Pi)_{\mathcal{C}}$. That is, we must show $F(\Pi)_{\mathcal{C}}\, M(\Pi) \subset M(\Pi)_{\mathcal{C}}$. We check it on generators. Let $x \in F\Pi'$ and $m \in M(\Pi)$. Then

$(i_*x)\cdot m = i_*(x\cdot i^*m) \in M(\Pi)_{\mathcal{C}}$. Done.

Lemma 2.7. Let R be a ring and N a module over R. If there exists an integer n such that $nx = 0$ for all $x \in R$, then n annihilates N.

Proof. $nx = n(1x) = (n1)x = 0x = 0$.

Corollary 2.8. If $F(\Pi)/F(\Pi)_{\mathcal{C}}$ has exp|n so does $M(\Pi)/M(\Pi)_{\mathcal{C}}$.

Corollary 2.9. Let A be an algebra over R. Then $R \to A$ induces $G_0^R(R\Pi) \to G_0^A(A\Pi)$ and makes $G_0^A(A\Pi)$ a $G_0^R(R\Pi)$ module. So if $G_0^R(R\Pi)/G_0^R(R\Pi)_{\mathcal{C}}$ has exp|n, so does $G_0^A(A\Pi)/G_0^A(A\Pi)_{\mathcal{C}}$.

If we could prove Artin's theorem for Z we would have it for any ring, since any ring is cannonically an algebra over Z. Artin's theorem is true for Z with n replaced by n^2. We prove this later.

We can dualize the above theory using i^*. Let M be a module over the Frobenius functor F and $\mathcal{C}$ be a class of groups.

Definition. $$M(\Pi)^{\mathcal{C}} = \bigcap_{\substack{\Pi' \subset \Pi \\ \Pi' \in \mathcal{C}}} \ker[M(\Pi) \xrightarrow{i^*} M(\Pi')].$$

Lemma 2.10. a) $M(\Pi)^{\mathcal{C}}$ is a $F(\Pi)$ submodule of $M(\Pi)$,

b) $M(\Pi)^{\mathcal{C}}$ is a module over $F(\Pi)/F(\Pi)_{\mathcal{C}}$.

c) If $F(\Pi)/F(\Pi)_{\mathcal{C}}$ has exp|n so does $M(\Pi)^{\mathcal{C}}$.

Proof. a) and c) are clear.

b) Observe that we only need $F(\Pi)_{\mathcal{C}} \cdot M(\Pi)^{\mathcal{C}} = 0$. If $x \in F(\Pi')$ with $\Pi' \in \mathcal{C}$ and $y \in M(\Pi)^{\mathcal{C}}$, then $(i_*x)\cdot y = i_*(x\cdot i^*y)$. But $i^*y = 0$. Hence $(i_*x)\cdot y = 0$.

Now we state an extension of Artin's theorem.

Theorem 2.11. Let Π be a finite group, $\mathcal{C}$ a class of subgroups, R a Dedekind ring, and K its quotient field. If $G_0(K\Pi)/G_0(K\Pi)_{\mathcal{C}}$ has exp|n then

a) if p is a nonzero prime ideal of R and k = R/p, then $G_0(k\Pi)/G_0(k\Pi)_{\mathcal{C}}$ has exp|n, and

b) $G_0(R\Pi)/G_0(R\Pi)_{\mathcal{C}}$ has exp$|n^2$.

Corollary 2.12. Let Π be a finite group of order n and $\mathcal{C}$ be the class of cyclic subgroups of Π, then

a) if R is an algebra of characteristic p, p ≠ 0, then

$G_0(R\Pi)/G_0(R\Pi)_{\mathcal{C}}$ has exp|n, and

b) if R is any ring, then $G_0^R(R\Pi)/G_0^R(R\Pi)_{\mathcal{C}}$ has exp$|n^2$.

Proof of corollary is easy from Artin's theorem, theorem 2.11 applied to Z and corollary 2.9.

These are not the best possible results. Lam [L] has improved on them by improving the exponent in Artin's theorem.

Proof of Theorem 2.11. By Theorem 1.9, there is a map $G_0(R\Pi) \to G_0(R/p\Pi)$ which is clearly a morphism of Frobenius functors since $G_0(R\Pi) \to G_0(K\Pi)$ and $G_0(K\Pi) \to G_0(R/p\Pi)$ are. Therefore $G_0(R/p\Pi)$ is a Frobenius module over $G_0(K\Pi)$. This proves (a) by Corollary 2.8.

For (b), we are given $x \in G_0(R\Pi)$ and must show $n^2x \in G_0(R\Pi)_{\mathcal{C}}$.

We examine the diagram

$$\begin{array}{ccccccc}
\coprod_{\substack{p \subset R \\ p \neq 0}} G_0(R/p\Pi) & \xrightarrow{g} & G_0(R\Pi) & \xrightarrow{f} & G_0(K\Pi) & \longrightarrow & 0 \\
\cup & & \cup & & \cup & & \\
\coprod_{\substack{p \subset R \\ p \neq 0}} G_0(R/p\Pi)_{\mathcal{C}} & \longrightarrow & G_0(R\Pi)_{\mathcal{C}} & \longrightarrow & G_0(K\Pi)_{\mathcal{C}} & \longrightarrow & 0
\end{array}$$

We have $f(nx) = nf(x) \in G_0(K\Pi)_{\mathcal{C}}$.
If $G_0(R\Pi)_{\mathcal{C}} \to G_0(K\Pi)_{\mathcal{C}}$ were onto then there would exist $y \in G_0(R\Pi)_{\mathcal{C}}$ with $f(y) = f(nx)$. Thus $f(nx-y) = 0$. Hence by exactness $nx-y = g(z)$ for some $z \in \coprod_{\substack{p \subset R \\ p \neq 0}} G_0(R/p\Pi)$. Now, by (a), $nz \in \coprod_{\substack{p \subset R \\ p \neq 0}} G_0(R/p\Pi)_{\mathcal{C}}$. Thus $n^2x = ny + g(nz)$ and $n^2x \in G_0(R\Pi)_{\mathcal{C}}$ as needed. It remains to check that $G_0(R\Pi)_{\mathcal{C}} \to G_0(K\Pi)_{\mathcal{C}}$ is onto. It is enough to show we can get a set of generators for $G_0(K\Pi)_{\mathcal{C}}$. Pick $y \in G_0(K\Pi')$ where $\Pi' \in \mathcal{C}$. Then there exists a $z \in G_0(R\Pi)$ mapping onto it. Clearly $i_*(z)$ maps onto $i_*(y)$. Done.

<u>Corollary 2.12</u>. Let Π be a finite group of order n and $\mathcal{C}$ the class of cyclic groups, then $G_0(Z\Pi)/G_0(Z\Pi)_{\mathcal{C}}$ has $\exp | n^2$.

<u>Corollary 2.13</u>. Let Π be a finite group of order n, $\mathcal{C}$ the class of cyclic groups and R any ring, then $G_0^R(R\Pi)/G_0^R(R\Pi)_{\mathcal{C}}$ and $K_0(R\Pi)/K_0(R\Pi)_{\mathcal{C}}$ has $\exp | n^2$.

Before we can prove Artin's Theorem we need some preliminary remarks on trace and class function.

Let K be a field of characteristic 0, A a finite dimensional K algebra, and M a finite dimensional A module. Pick $a \in A$ and let $\hat{a}: M \to M$ denote the function $\hat{a}(m) = am$. $X_M(a)$ = trace of $\hat{a}$ on M. The trace is K-linear in a, that is, if $a, b \in A$ and $r, s \in K$ then

$$X_M(ra + sb) = rX_M(a) + sX_M(b).$$

If $0 \to M' \to M \to M'' \to 0$ is an exact sequence of finite dimensional A module, then we can pick a base of M" and lift its elements to elements of M. The resulting set, together with a base for M' will be a base for M. With respect to that base the matrix for $\hat{a}$ on M looks like

$$\begin{pmatrix} \hat{a} \text{ on } M' & 0 \\ * & \hat{a} \text{ on } M'' \end{pmatrix}$$

Trace is diagonal sum, therefore, $X_M(a) = X_{M'}(a) + X_{M''}(a)$. Thus, $X(a)$: $G_0(A) \to K$ when we think of the trace as a function of the module. Since $X_M \in \mathrm{Hom}_K(A,K)$ and $X_M = X_{M'} + X_{M''}$ we can regard X as a map from $G_0(A)$ to $\mathrm{Hom}_K(A,K)$.

<u>Theorem 2.14</u>. If K is a field of characteristic 0 and A a finite dimensional K-algebra, then X: $G_0(A) \to \mathrm{Hom}_K(A,K)$ is injective.

<u>Remark</u>. $G_0(A)$ is free abelian on $[S_i]$ where S_i are representatives of the isomorphism classes of simple modules. Since $\mathrm{Hom}_K(A,K)$ is a divisible group for char K = 0, X cannot be onto. Also, if K has characteristic $p \neq 0$, X cannot be injective.

<u>Proof</u>. Pick $x \in G_0(A)$. Then $x = \sum m_i[S_i] = [M] - [N]$ where M and N are semisimple A modules. Let $\bar{A} = A/\mathrm{rad}(A)$. Then M and N are $\bar{A}$ modules. Thus we need to show that if M and N are semisimple A module and $X_M = X_N$, then M is isomorphic to N.

$M = \coprod_i (\coprod_1^{n_i} S_i)$. If we can determine the n_i we are done. $A = \prod_{i=1}^{r} A_i$ where the A_i are simple rings and the S_i are simple A_i modules. Let e_i be the identity of A_i. Then the e_i are primitive central idempotents. $\hat{e}_i: S_j \to S_j$ is $\delta_{i_j} 1_{S_j}$ where δ_{i_j} is the Kronecher delta. Hence, $\hat{e}_j: M \to M$ has a matrix which looks like $\begin{pmatrix} I & 0 \\ 0 & 0 \end{pmatrix}$ if we put the S_j's first. Hence the trace is $n_j \dim S_j$.

Pick $a_i \in A$ mapping onto $e_i \in \bar{A}$. Then $X_M(a_i) = n_i \dim S_i$ since a_i acts on M through e_i. $\dim S_i \neq 0$ so we can solve for n_i uniquely. Note that we need characteristic 0 here. Thus we can recover the n_i from M. Thus $X: G_0(A) \to \mathrm{Hom}_K(A,K)$ is one-one.

Done.

Let $A = K\Pi$ where Π is a finite group. $\mathrm{Hom}_K(A,K) = \{$function from Π to $K\}$. $X: G_0(K\Pi) \to \{$K valued functions on $\Pi\}$. We make the latter a ring by $(fg)(s) = f(s)g(s)$ for all $s \in \Pi$.

<u>Theorem 2.15</u>. X is a ring homomorphism.

<u>Proof</u>. Let M and N be $K\Pi$ modules and $s \in \Pi$. We need to calculate $X_{M \otimes N}(s)$. We need the trace of $\hat{s}: M \otimes N \to M \otimes N$. $\hat{s}(m \otimes n) = sm \otimes sn$. If m_i is a base for M and n_j a base for N then $M \otimes N$ has $m_i \otimes n_j$. If $sm_i = \sum a_{iK} m_K$ and $sn_j = \sum b_{je} n_j$, then

$s(m_i \otimes n_j) = sm_i \otimes sn_j = \sum_{k,\chi} a_{ik}b_{j\chi}(m_k \otimes n_\chi)$. So the trace is $\sum_{i,j} a_{ii}b_{jj} = (\sum a_{ii})(\sum b_{jj})$. Thus, $X_{M \otimes N}(s) = X_M(s)X_N(s)$.

Next we check that the map preserves the identity. [K] is the unit of $G_0(K\Pi)$. $s1 = 1$ for all $s \in \Pi$. Thus $X_K(s) = 1$.

Done.

Thus X identifies $G_0(K\Pi)$ with a subring of K valued functions on Π.

Now let $F(\Pi, K)$ be the ring of K valued functions on Π. If $i: \Pi' \to \Pi$ are homomorphisms of groups, we clearly get a commutative diagram

$$\begin{array}{ccc} G_0(K\Pi) & \xrightarrow{i^*} & G_0(K\Pi') \\ \downarrow X & & \downarrow X \\ F(\Pi,K) & \xrightarrow{i^*} & F(\Pi',K) \\ f & \rightsquigarrow & fi \end{array}$$

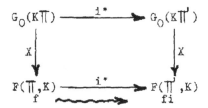

If i is a monomorphism (which we can assume is an inclusion), then we have $\Pi = \cup s_i\Pi'$ where the s_i are coset representatives. Then $K\Pi$ is free as a right $K\Pi'$ module with s_i as a base and $K\Pi \otimes_{K\Pi'} M = \coprod(s_i \otimes M)$. That is, every element is uniquely expressible in the form $\sum s_i \otimes m_i \quad m_i \in M$. If $t \in \Pi$, then $t(\sum s_i \otimes m_i) = \sum ts_i \otimes m_i$ is not in the correct form. However, $\cup ts_i\Pi' = t\Pi = \Pi$ and the union is disjoint. Hence there is a permutation of the cosets induced by t such that $ts_i\Pi' = s_{t(i)}\Pi'$. That is, there are elements $p_{t,i} \in \Pi'$ such that $ts_i = s_{t(i)}p_{t,i}$. Therefore, $\sum ts_i \otimes m_i = \sum s_{t(i)}p_{t,i} \otimes m_i = \sum s_{t(i)} \otimes p_{t,i}m_i$. Let $a_1, \ldots, a_n$ be a base for M over K. Then $K\Pi \otimes_{K\Pi'} M$ has a base $\{s_i \otimes a_j\}$. If $p \in \Pi'$, then $pa_j = \sum_k R(p)_{j,k}a_k$ with $R(p) \in K$. Thus, $s_{t(i)} \otimes p_{t,i}a_j = \sum_k s_{t(i)} \otimes R(p_{t,i})_{j,k}a_k = \sum_k R(p_{t,i})_{j,k}(s_{t(i)} \otimes a_k)$. That is, the coefficient of $s_i \otimes a_j$ in above is 0 if $t(i) \neq i$ and $R(p_{t,i})_{j,j}$ if $t(i) = i$. Thus we can compute the trace by

$$X_{K\Pi \otimes_{K\Pi'} M}(t) = \sum_{\substack{i \text{ such that} \\ t(i) = i}} \sum_j R(p_{t,i})_{j,j}$$

$$= \sum_{\substack{i \text{ such that} \\ t(i) = i}} \sum_j \text{(trace of } p_{t,i} \text{ on M)}.$$

That is,

$$X_{i_*(M)}(t) = \sum_{\substack{i \text{ such that} \\ t(i) = i}} X_M(p_{t,i}).$$

$t(i) = i$ if and only if $ts_i = s_i p_{t,i}$ with $p_{t,i} \in \Pi'$ which happens if and only if $s_i^{-1}ts_i = p_{t,i} \in \Pi'$. If we define

$$X_M(p) = \begin{cases} X_M(p) & \text{if } p \in \Pi' \\ 0 & \text{if } p \notin \Pi' \end{cases}, \text{ then } X_{i_*(M)}(t) = \sum_{s \in \Pi} X_M(s^{-1}ts).$$

<u>Lemma 2.16</u>. Let K be any field, Π any finite group and M a finite $K\Pi$ module, then $X_M(t) = X_M(sts^{-1})$ for all $s,t \in \Pi$.

<u>Proof</u>. Consider the commutative diagram where the columns are isomorphisms

$$\begin{array}{ccc} M & \xrightarrow{\hat{t}} & M \\ \hat{s}\downarrow & & \downarrow\hat{s} \\ M & \xrightarrow{\widehat{sts-1}} & M \end{array}$$

Therefore, trace $\hat{t}$: $M \to M$ is the same as trace $\widehat{sts^{-1}}$: $M \to M$.

Done.

Now we return to $\Pi = \cup s_i\Pi'$. Pick $p \in \Pi'$, then $(s_i^{-1}ts_i) = X(p^{-1}s_i^{-1}ts_ip)$ since one of the products is in Π' if and only if the other is. Hence $X(s_i^{-1}ts_i) = \frac{1}{|\Pi'|}\sum_{p\in\Pi'} X(p^{-1}s_i^{-1}ts_ip)$ where $|\Pi'|$ is the order of Π'. Finally

$$X_{i_*(M)}(t) = \frac{1}{|\Pi'|}\sum_{s\in\Pi} X_M(s^{-1}ts).$$

<u>Definition</u>. A function $f: \Pi \to K$ is a <u>class</u> <u>function</u> if $s = ptp^{-1}$ for some $t \in \Pi$ implies $f(s) = f(t)$. Let $F_a(\Pi,K)$ be the set of such functions.

Generalizing the formula for $X_{i_*(M)}$ just given, we can define i_*: $F_a(\Pi',K) \to F_a(\Pi,K)$ such that the diagram

$$\begin{array}{ccc} G_0(K\pi') & \xrightarrow{i_*} & G_0(K\pi) \\ \downarrow X & & \downarrow X \\ F_a(\pi',K) & \xrightarrow{i_*} & F_a(\pi,K) \end{array}$$ commutes.

<u>Theorem 2.17</u>. Let M be a $Q\pi$ module where π is a finite group. If $s,t \in \pi$ generate the same subgroup ($\langle s\rangle = \langle t\rangle$), then $X_M(s) = X_M(t)$.

<u>Remark</u>. This is false if Q is replaced by $\mathbb{C}$, the complex numbers.

<u>Proof</u>. It is enough to consider the case $\pi' = \langle s\rangle = \langle t\rangle$. M as a $Q\pi'$ module has trace equal to X_M restricted to π' ($X_M|_{\pi'}$). Therefore, it is enough to consider the case of π cyclic of order n.

$Q\pi = Q[x]/(x^n-1)$ where $x \rightsquigarrow$ generator of π. Let $\Phi_n(x) = \prod_{\substack{\text{primitive} \\ \text{n-th roots} \\ \text{of } 1}} (x-r_i) \in Q[x]$. Then $x^n-1 = \prod_{\substack{\text{all n-th} \\ \text{roots of} \\ 1}} (x-r_i) = \prod_{d/n} \Phi_d(x)$.

Since the Φ_d are relatively prime (having no common factor $x-r_i$), we have $Q\pi \cong \prod_{d|n} Q[x]/(\Phi_d(x))$.

We use the fact that $\Phi_d(x)$ are irreducible over Q and note that the remainder of this proof works for any field such that all $\Phi_d(x)$ are irreducible.

$$Q\pi = Q[x]/(x^n-1) = \prod Q[x]/\Phi_d(x) = \prod K_d$$

where K_d is Q with a primitive d-th root r_d of 1 adjoined. All K_d are fields. We only need to check the theorem for simple modules. The simple modules are just K_d. π acts on K_d through x by

$\hat{r}_d: K_d \to K_d$ giving the usual trace $T_{Q(r_d)/Q}(r_d) \in Q$ and

$T_{Q(r_d)/Q} = \sum$ conjugates of r_d. Hence $T_{Q(r_d)/Q}$ is independent of which generator of Π was picked since the conjugates are all primative d-th roots of 1. Therefore $X(s) = X(t)$. Done.

Now let K be any field of characteristic 0. First we show that for any $K\Pi$ module M and any $s \in \Pi$ that $X_M(s)$ is a sum of roots of unity. Let L be a field containing K. Then $X_{L \otimes_K M}(s) = X_M(s)$ because if e_i is a K base for M then $1 \otimes e_i$ is a L base for $L \otimes_K M$. Then $\hat{s}$ has the same matrix for both $L \otimes_K M$ and M and hence the characters are the same. Thus we can assume K is algebraically closed to compute characters. Then $x^n - 1 = \prod_{\substack{\text{all } n\text{-th} \\ \text{roots of } 1, r_i}} (x - r_i)$.

$K\langle s \rangle = K[x]/(x^n-1) = \prod_1^n K_{r_i}$ and s acts on K_{r_i} by multiplication by r_i.

<u>Theorem 2.18</u>. Let R be an integrally closed domain of characteristic 0, X be a trace of a $K\Pi$ module where K is the quotient field of R. Then $X(s) \in R$ for all $s \in \Pi$.

<u>Proof</u>. $X(s)$ is a sum of roots of unity and hence is integral over R. $X(s) \in K$ by definition. But R integrally closed so $X(s) \in R$.

Done.

<u>Definition</u>. Let Π be a group and $s, t \in \Pi$. Write $s \sim_Q t$ if $\langle s \rangle$ and $\langle t \rangle$ are conjugate subgroups.

We can summarize what we have done so far as follows: If X is a character of a $Q\pi$ module then 1) $s \sim_Q t$ implies $X(s) = X(t)$ and $X(s) \in Z$ all $s \in \pi$.

<u>Theorem 2.19</u>. (Artin). Let π be a finite group of order n and f be a function on π such that

a) $f(s) \in Z$ for all $s \in \pi$ and

b) $s \sim_Q t$ implies $f(s) = f(t)$.

Then $nf = \sum a_v i_{v*}(1_{\pi_v})$ where π_v runs over the cyclic subgroups of π, 1_{π_v} is the character of the trivial representation, and $a_v \in Z$.

<u>Remark</u>. If $\mathcal{C}$ is the class of cyclic groups, apply theorem 2.19 to the function f which is identically 1 and get $nf \in G_0(Q\pi)_{\mathcal{C}}$ and hence that $G_0(Q\pi)/G_0(Q\pi)_{\mathcal{C}}$ has $\exp | n$. That is, this will complete the proof of theorem 2.5.

<u>Proof</u>.

1) It is enough to prove this for the case where f is the characteristic function of a Q class that is $\pi = \cup K_i$ where each K_i is the equivalence class of the equivalence relation $\sim_Q$. Let $f_i(s) = \begin{cases} 1 & s \in K_i \\ 0 & s \notin K_i \end{cases}$. Then any $f = \sum f(s_i) f_i$ where s_i runs over a set of representatives of the classes.

2) $i_*: \pi' \to \pi$ is an inclusion of finite groups and g is a function on π' of the form $g = \sum a_v i_{v*}(1_{\pi_v})$ over $\pi_v \subset \pi$ cyclic and

$a_v \in Z$ then $i_*(g) = i_*(\sum a_v i_{v_*}(1_{\Pi_v})) = \sum a_v i_* i_{v_*}(1_{\Pi_v}) =$

$= \sum a_v (ii_{v_*})(1_{\Pi_v})$ is of the same form.

We finish the proof by an induction on the order of Π. Let f be the characteristic function of the Q-class of $s \in \Pi$.

<u>Case 1</u>. $s \in \Pi$ does not generate Π. $\Pi' = \langle s \rangle$. By induction the theorem is true for Π'. Now the function

$g(x) = \begin{cases} 1 & \text{if x generates } \Pi' \\ 0 & \text{if x does not generate } \Pi' \end{cases}$, is the characteristic function of the Q class of s in Π'. If $n' = |\Pi'|$, then we know $n'g(x) = \sum a_v i_{v_*}(1_{\Pi_v})$. By 2) $i_*(n'g)$ has the required form.

$$\begin{aligned} i_*(n'g)(t) &= \frac{1}{|\Pi'|} \sum_{p \in \Pi} n'g(ptp^{-1}) \\ &= \sum_{p \in \Pi} g(ptp^{-1}) \\ &= \sum_{p \in \Pi} \begin{cases} 1 \text{ if } ptp^{-1} \text{ generates } \Pi' \\ 0 \text{ if not} \end{cases} \\ &= \text{the number of p for which } ptp^{-1} \\ &\quad \text{generates } \Pi' \end{aligned}$$

If t is not Q equivalent to s this number is 0. If $t \sim_Q s$ $\langle t \rangle = \langle asa^{-1} \rangle = a\Pi' a^{-1}$. Now ptp^{-1} generates

$p\langle t\rangle p^{-1} = p a \pi' a^{-1} p^{-1}$. That is, ptp^{-1} generates π' if and only if $pa \in N(\pi')$, the normalizer of π'. This happens if and only if $p \in N(\pi)a^{-1}$. The set $N(\pi)a^{-1}$ has $|N(\pi)|$ elements.

If f is the characteristic function of the Q class of s, then $i_*(n'g) = |N(\pi')|f$. Therefore, $nf = [\pi: N(\pi')]i_*(n'g)$ and nf has the required form.

Case 2. $s \in \pi$ generates π and hence π is cyclic and the characteristic function f of the Q class of s is

$$f(t) = \begin{cases} 1 & \text{if } t \text{ generates } \pi \\ 0 & \text{if not} \end{cases}$$

π itself is a cyclic subgroup of π. $1_\pi = i_*(1_\pi)$ where $i: \pi \to \pi$ is the identity. Therefore, if $n(1_\pi - f)$ has the required form so does f.

$$(1_\pi - f)(t) = \begin{cases} 0 & \text{if } t \text{ generates } \pi \\ 1 & \text{if not.} \end{cases}$$

Let K_1 be the Q class of s and $K_2,\ldots,K_r$ be the remaining Q classes. Then $1_\pi - f = \sum$ characteristic function of $K_2,\ldots,K_r$. But these functions have the desired form by Case 1. This completes the induction step and the proof.

This gives information on the rank of $G_0(Q\pi)$. If $F_{Qcl}(\pi,Z) = \{f \in F(\pi,Z) \mid x \sim_Q y \Rightarrow f(x) = f(y)\}$, then $F_{Qcl}(\pi,Z) \supset$ Character ring $\supset nF_{Qcl}(\pi,Z)$. The character ring and $G_0(Q\pi)$ can be identified. Therefore, rank $G_0(Q\pi)$ equals the number of irreducible $Q\pi$ modules equals the number of Q classes of π.

Theorem 2.20. If k is a field, then the Cartan map $\kappa: K_0(k\pi) \to G_0(k\pi)$ is a monomorphism and the coker is finite.

Remark. Both sides are finitely generated free abelian groups.

Proof. $K_0(k\pi)$ and $G_0(k\pi)$ are Frobenius modules and κ is a Frobenius homomorphism. Therefore the kernel of κ and coker κ are Frobenius modules. It is enough to prove the theorem in the case of π finite cyclic. Say $\pi = \langle x \rangle$ where $x^n = 1$. Then $k\pi = k[x]/(x^n-1)$. Let $x^n-1 = \prod f_i(x)^{e_i}$ be the factoring of x^n-1 into powers of distinct irreducibles. Then $k\pi = \prod k[x]/(f_i(x)^{e_i})$. Thus, $k\pi/\mathrm{rad}(k\pi) = \prod k[x]/(f_i(x))$. The simple modules are $S_i = k[x]/(f_i(x))$. If f is irreducible, then $A = k[x]/(f(x)^e)$ is a local ring and $A/(f(x))$ is a field. A is projective, being a direct summand of $k\pi$ and is indecomposable by Nakayama's lemma. Let $P_i = k[x]/(f_i(x)^{e_i})$. Then the $[P_i]$ are a base for $K_0(k\pi)$ and the S_i are a base for $G_0(k\pi)$. We can filter P_i as $P_i \supset (f_i(x)) \supset \dots \supset (f_i(x)^{e_i}) = 0$ with successive quotients S_i. Therefore $\kappa[P_i] = e_i[S_i]$ and κ is a monomorphism. The coker κ has order $\prod e_i$ and hence is finite. Done.

Theorem 2.21. If R is a local domain with quotient field K, π is a finite group, and P and Q are finitely generated projective $R\pi$ modules with $K \otimes_R P$ isomorphic to $K \otimes_R Q$, then P is isomorphic to Q.

Proof. Immediate from Theorem 2.20 and Theorem 1.10.

Chapter 3: Finiteness Theorems

Let R be a Dedekind ring, A an R algebra which is finitely generated as an R module, and let M and N be A modules which are finitely generated projective R modules.

Definition. If M and N are as above, we say that they have the same genus if M_p is isomorphic to N_p over A_p for all prime ideals p of R.

Theorem 2.20 asserts that if P and Q are finitely generated projective $R\pi$ modules with $K \otimes_R P$ isomorphic to $K \otimes_R Q$ where K is the quotient field of R, then P and Q have the same genus.

Theorem 3.1. (Roiter). If M and N have the same genus and I is a nonzero ideal of R, then there exists an exact sequence $0 \to M \to N \to X \to 0$ with the R annihilator of X prime to I.

We use without proof the following well known

Proposition 3.2. Let R be a commutative ring, S a multiplicatively closed subset of R, A an R algebra, M and N A modules with M finitely presented then $\mathrm{Hom}_{A_S}(M_S, N_S)$ is naturally isomorphic to $(\mathrm{Hom}_A(M, N))_S$.

Proof of Theorem 3.1. Let $p_1, \ldots, p_n$ be $n \geq 1$ be a finite set of primes of R including all the primes that contain I. Let $g_i: M_{p_i} \to N_{p_i}$ be any isomorphism. By Proposition 3.2 $g_i = f_i/s_i$ where $s_i \in R - p_i$ and $f_i \in \mathrm{Hom}_A(M, N)$. $f_i: M_{p_i} \to N_{p_i}$ is also an isomorphism. By the Chinese Remainder Theorem there exist $a_i \in R$

such that $a_i = \begin{cases} 1 & \text{mod } p_i \\ 0 & \text{mod } p_j \quad i \neq j \end{cases}$. Let $f = \sum_{i=1}^{n} a_i f_i$. Then

$f: M_{p_i} \to N_{p_i}$ $i = 1, \ldots, n$ is an isomorphism. For

$f - f_i = \sum_{j \neq i} a_j f_j + (a_i - 1) f_i$ is an element of $p_i \mathrm{Hom}_A(M, N)$.

Therefore, $f - f_i : M \to p_i N$. Thus locally $f - f_i$: $M_{p_i} \to p_i N_{p_i}$. Reducing mod p_i shows that $\bar{f} - \bar{f}_i$: $M_{p_i}/p_i M_{p_i} \to N_{p_i}/p_i N_{p_i}$ is the 0 map. Therefore $\bar{f}$ and $\bar{f}_i$ agree as maps $M_{p_i}/p_i M_{p_i} \to N_{p_i}/p_i N_{p_i}$. But $\bar{f}_i$ is an isomorphism and therefore $\bar{f}$ is. By Nakayama's lemma $f: M_{p_i} \to N_{p_i}$ is onto. Forming the exact sequence $0 \to Y \to M \xrightarrow{f} N \to X \to 0$ we know that $X_{p_i} = 0$ and hence the R annihilator of X is not contained in p_i $i = 1, \ldots, n$. Tensoring with K we get the exact sequence

$$0 \to K \otimes Y \to K \otimes M \xrightarrow{1 \otimes f} K \otimes N \to K \otimes X \to 0.$$

But $K \otimes X = 0$ and $K \otimes M$ and $K \otimes N$ have the same dimension as vector spaces. Therefore, $1 \otimes f$ is an isomorphism. Thus, Y is torsion but M is torsion free since it is projective. Thus $Y = 0$.

Done.

Theorem 3.3. Let R be a Dedekind ring with quotient field K, π a finite group whose order n is not divisible by the characteristic of K. Let P be a finitely generated projective $R\pi$ module with

$K \otimes_R P$ free on m generators as a $K\pi$ module, then P is isomorphic to $R\pi \oplus \ldots \oplus R\pi \oplus I$ where I is an ideal of $R\pi$, and if $\mathfrak{A}$ is a nonzero ideal of R, then $\mathrm{ann}_R(R\pi/I)$ can be chosen prime to $\mathfrak{A}$.

<u>Proof</u>. Pick an ideal $\mathfrak{A} \neq 0$, then $n\mathfrak{A} \neq 0$. Apply Theorem 3.1 and the remark preceding it to obtain $0 \to P \to F \to X \to 0$ an exact sequence with annihilator X prime to $n\mathfrak{A}$ and F free on m generators. Let pr_m be the projection of $F = R\pi \oplus \ldots \oplus R\pi$ on the m-th factor. pr_m restricted to P maps P onto an ideal I of $R\pi$. The annihilator of $R\pi/I$ is prime to $n\mathfrak{A}$ for the diagram

$$\begin{array}{ccccccccc} 0 & \to & P & \to & F & \to & X & \to & 0 \\ & & \downarrow & & \downarrow & & \downarrow & & \\ 0 & \to & I & \to & R\pi & \to & Y & \to & 0 \\ & & \downarrow & & \downarrow & & & & \\ & & 0 & & 0 & & & & \end{array}$$

has rows and columns exact and commutes.

Therefore, $X \to Y$ is onto. Thus the annihilator of Y contains the annihilator of X and hence is prime to $n\mathfrak{A}$. Next we need two lemmas.

<u>Lemma 3.4</u> (Rim). Let R be a Dedekind ring and π a finite group of order n. If Y is any $R\pi$ module and the R annihilator of Y is prime to n, then the projective dimension of Y over $R\pi$ ($pd_{R\pi} Y$) is ≤ 1.

<u>Proof</u>.

1) $\hat{n}: Y \longrightarrow Y$, $x \rightsquigarrow nx$ is an isomorphism since its kernel and cokernel are annihilated by both n and the R-annihilator of Y.

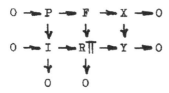

2) Let M be an $R\Pi$ module, $f: M \to M$ an R homomorphism, then $tr(f)$ defined by $tr(f)(m) = \sum_{s \in \Pi} sf(s^{-1}m)$ is an $R\Pi$ homomorphism.

Let $t \in \Pi$ then

$$\begin{aligned} tr(f)(tm) &= \sum_{s \in \Pi} sf(s^{-1}tm) \qquad \text{Letting } p = t^{-1}s \text{ yields} \\ &= \sum_{p \in \Pi} tpf(p^{-1}m) \\ &= t\sum_{p \in \Pi} pf(p^{-1}m) \\ &= t\ tr(f)m. \end{aligned}$$

3) If f is an $R\Pi$ homomorphism, then

$tr(f)(m) = \sum_{s \in \Pi} sf(s^{-1}m) = \sum_{s \in \Pi} f(m) = nf(m)$. Hence if $\hat{n}: X \to X$ is an isomorphism, its inverse g is an $R\Pi$ map so $tr(g)(x) = ng(x) = x$. We have produced an $R\Pi$ map whose trace is the identity.

4) Next we prove that if R is Dedekind, Π finite, X an $R\Pi$ module with an R homomorphism $g: X \to X$ whose trace is 1_X, then $pd_{R\Pi} X \leq 1$. This will complete the lemma.

<u>Definition</u>. Let R be any ring, Π a finite group, and P an $R\Pi$ module. Then P is <u>relatively projective</u> (relative to R) if for every diagram of $R\Pi$ modules

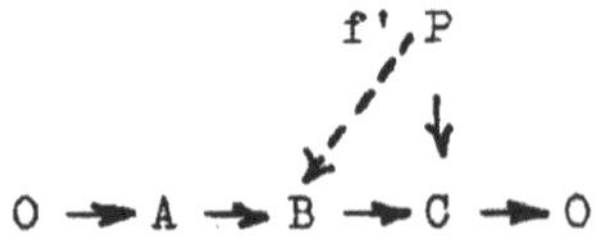

with exact row where f is defined over R and makes the diagram commute, there is a map f: P $\to$ B defined over RΠ making the diagram commute.

We observe that the existence of the function g in 3) proves that X is relatively projective. Suppose we have a diagram

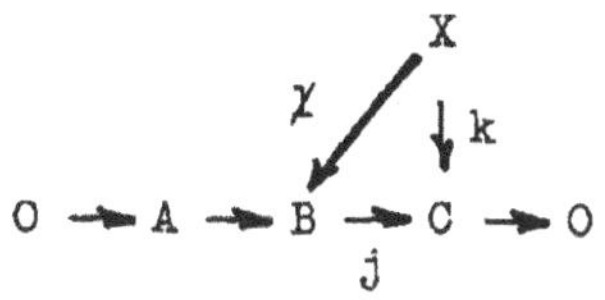

where χ is an R map.

Define h: X $\to$ B by h = tr(χg). Then h is an RΠ map and jh = k.

$$\begin{aligned} jh(x) &= j(\textstyle\sum s\chi g(s^{-1}x)) \\ &= \textstyle\sum sj\chi g(s^{-1}x) \\ &= \textstyle\sum skg(s^{-1}x) \\ &= k(\textstyle\sum sg(s^{-1}x)) \\ &= k(tr(g)(x)) \\ &= k(x). \end{aligned}$$

Next we look at the RΠ map

$$R\Pi \otimes_R X \to X \qquad \text{given by}$$

$s \otimes x \leadsto sx$ for all $s \in \Pi$ $\quad x \in X$. This is clearly onto with kernel Y. We have a diagram

$$\begin{array}{ccccccccc} & & & & & & X & & \\ & & & & & \overset{f}{\swarrow} & \downarrow 1_X & & \\ 0 & \to & Y & \to & R\pi \otimes_R X & \to & X & \to & 0 \end{array}$$

where f is the R map $x \rightsquigarrow 1 \otimes x$. If X is relatively projective, this splits over $R\pi$, and, therefore, X is a direct summand of $R\pi \otimes_R X$. Thus $\mathrm{pd}_{R\pi} X \leq \mathrm{pd}_{R\pi}(R\pi \otimes_R X)$. We regard X as an R module. Then, since R is Dedekind, we have a resolution

$$0 \to P \to F \to X \to 0 \quad \text{where } F \text{ is free}$$

and P is projective over R. Thus $\mathrm{pd}_R X \leq 1$. Tensoring with $R\pi$ (which is exact) gives

$$0 \to R\pi \otimes_R P \to R\pi \otimes_R F \to R\pi \otimes_R X \to 0$$

which shows $\mathrm{pd}_{R\pi}(R\pi \otimes_R X) \leq 1$. This completes Rim's lemma.

<u>Lemma 3.5</u>. Let I and J be projective ideals of $R\pi$ such that $KI = KJ = K\pi$. Then $I \oplus J$ is isomorphic to $R\pi \oplus L$ where L is a (projective) ideal of $R\pi$.

<u>Proof</u>. By Theorem 3.1 I is isomorphic to an ideal I_1 so that $\mathrm{ann}(R\pi/I_1) = \mathfrak{a}$ is prime to $\mathrm{ann}(R\pi/J)$. Let $I_1 \oplus J \to R\pi \to X \to 0$ be the map induced from the inclusion of I_1, J in $R\pi$. Then $\mathrm{ann}(X) \supset \mathfrak{a}$ and $\mathrm{ann}(X) \supset \mathrm{ann}(R\pi/J)$ which is prime to $\mathfrak{a}$. Therefore X is 0 and we get a splitting $0 \to L \to I_1 \oplus J \leftrightarrows R\pi \to 0$. This completes the lemma.

Returning to the proof of the theorem we observe in the diagram * that I is projective by Rim's lemma. Thus $P = I \oplus P'$.

Continuing with P' we get that $P' = I_1 \oplus P''$. Finally $P = I \oplus I_1 \oplus \dots \oplus I_r$. Then applying lemma 3.5 and induction we are done.

Corollary 3.6. If R is a field whose characteristic is prime to the order of a finite group π, then all modules are projective.

Proof. We need only show all short exact sequences split. Let $0 \to A \to B \to C \to 0$ be exact. Then there is a splitting $f: C \to B$ as R modules. $\frac{1}{n}tr(f)$ gives a splitting as $R\pi$ modules.

Definition. Let R be a domain with quotient field K and let A be a finite dimensional K algebra with 1. An R-order in A is a sub-R-algebra Λ of A which is finitely generated as an R-module and such that $K\Lambda = A$. These exist for all A by the proof of Lemma 3.7 (2) below with $R_s = K$ and $\Lambda = A$.

Definition. Let R be a domain with quotient field K. We say the Jordan-Zassenhaus theorem holds for R if for any finite dimensional semisimple K-algebra A, order Λ in A, and integer s, there are only finitely many isomorphism classes of Λ-modules M which are finitely generated and torsion free over R of rank $\leq$ s.

The classical Jordan-Zassenhaus asserts that R = Z has this property. We will prove this below. Further examples are given by the following lemma.

Lemma 3.7 (1) If $R \subset R'$ are domains with R' finitely generated as an R-module, and if the Jordan-Zassenhaus theorem holds for R, then it holds for R'.

(2) If S is a multiplicative set in R and if the Jordan-Zassenhaus theorem holds for R, then it holds for R_S.

Proof. (1) Let K' be the quotient field of R'. Then $K' \supset KR' \supset K$ and KR' is a domain finite dimensional over K. Therefore KR' is a field so K' = KR' and so $[K':K] < \infty$. If A is a finite dimensional K' algebra, it is finite dimensional over K. Any R'-order Λ in A is also an R order. If M is a Λ module finitely generated and torsion free of rank $\leq$ s over R', then Λ is finitely generated, torsion free of rank $\leq s[K':K]$ over R so there are only a finite number of possible isomorphism classes for M.

(2) Let A be a finite dimensional semisimple K-algebra and Λ an R_S-order in A. Let $\lambda_1, \ldots, \lambda_n \in \Lambda$ generate Λ as an R_S-module. Let $\lambda_i \lambda_j = \sum_k c_{ijk} \lambda_k$, $c_{ijk} \in R_S$. Replacing all λ_i by $s\lambda_i$, $s \in S$ changes c_{ijk} to sc_{ijk}. Therefore we can assume all $c_{ijk} \in R$. Let Γ be the R-submodule of Λ generated by $1, \lambda_1, \ldots, \lambda_n$. Then Γ is an R-order in A and $R_S\Gamma = \Gamma_S = \Lambda$. Now if M is a Λ-module, finitely generated and torsion free over R of rank $\leq$ s, let $m_1, \ldots, m_r$ generate M over Λ and let N be the Γ submodule generated by $m_1, \ldots, m_r$. Then N is finitely generated over Γ and torsion free over R of rank $\leq$ s so there are only a finite number of possibilities for N up to isomorphism. But N determines M since KN = KM and $M = R_S N = N_S \subset KN$.

Once we have proved the Jordan-Zassenhaus theorem for Z, this lemma shows that it holds for the ring R of integers in any algebraic number field and also for all rings of quotients R_S.

Before proving this, we give the application to K-theory.

Theorem 3.8. Let R be a Dedekind ring for which the Jordan-Zassenhaus theorem holds. Let K be the quotient field of R. Let π be a finite group of order prime to the characteristic of R. Then $K_0(R\pi)$ and $G_0(R\pi)$ are finitely generated abelian groups and the map $K_0(R\pi) \to G_0(K\pi)$ has a finite kernel, $C_0(R\pi)$.

Proof. $G_0(R\pi) = G_0{}^R(R\pi)$ is generated by $\{[M]\}$ where M is a finitely generated torsion free R module. Then $K \otimes_R M = W$ is a finitely generated $K\pi$ module. Therefore, W has a composition series, say

$$W = W_0 \supset W_1 \supset \ldots \supset W_r = 0 .$$

Let $M_i = W_i \cap M$. Then $M_i/M_{i+1} \subset W_i/W_{i+1}$ and, therefore, M_i/M_{i+1} is torsion free and is a finitely generated R-module since M is. Furthermore $rk(M_i/M_{i+1}) \leq \dim_K(W_i/W_{i+1})$. This dim is $\leq$ max of dim of simple $K\pi$ modules $\leq |\pi|$. But $[M] = \sum [M_i/M_{i+1}]$. Thus far we have shown that for any Dedekind ring R, $G_0(R\pi)$ is generated by [M] where M is finitely generated and torsion free as an R module and $K \otimes_R M$ is simple as $K\pi$ module. The Jordan-Zassenhaus theorem implies there are only finitely many isomorphism classes of such M. Hence $G_0(R\pi)$ is finitely generated. Since for any multiplication system $S \subset R, G_0(R_S\pi)$ is an image of $G_0(R\pi)$, $G_0(R_S\pi)$ is finitely generated also.

$G_0(K\pi)$ is a finitely generated free abelian group. Hence, if $\ker(K_0(R\pi) \to G_0(K\pi))$ is finite, then $K_0(R\pi)$ is finitely generated. Say $[P]-[Q] \rightsquigarrow 0$ in $G_0(K\Pi)$ then $[K \otimes_R P] = [K \otimes_R Q]$ and $K \otimes_R P$ is isomorphic to $K \otimes_R Q$. Find Q' such that $Q \oplus Q' = F$ is a finitely generated free $R\Pi$ module. Then $[P] - [Q] = [P \oplus Q'] - [Q \oplus Q'] = [P \oplus Q] - [F]$. So the kernel is generated by all $[P] - [F]$ where F is free and $K \otimes_R P$ is isomorphic to $K \otimes_R F$. Since $|\Pi|$ is prime to char K, Theorem 3.3 shows that $P = F' \oplus I$ and $F = F' \oplus R\Pi$ where F' is free and I is an ideal of $R\Pi$. Therefore, $[P] - [F] = [I] - [R\Pi]$ where I is an ideal such that $K \otimes_R I = K\Pi$. The Jordan-Zassenhaus Theorem implies there are only a finite number of isomorphism classes of such I. Therefore, the kernel is finite and $K_0(R\Pi)$ is finitely generated.

We now turn to the proof of the Jordan-Zassenhaus Theorem.

Theorem 3.9. Let R be Z or $k[X]$ where k is a finite field. Then the Jordan-Zassenhaus Theorem holds for R.

Corollary 3.10. If R is an order in an algebraic number field or an order over $k[X]$ in some finite extension of $k(X)$ k being a finite field, then the Jordan-Zassenhaus Theorem holds for R and for all rings of quotients R_S.

This follows from Lemma 3.7.

Proof of Theorem 3.9. The proof uses only the following 5 properties of R. (1) R is a principal ideal domain. (2) If $x \in R$, $x \neq 0$ then R/Rx is finite. Using (2) we define $|x|$ to be the order of R/Rx for $x \neq 0$ and set $|0| = 0$. Since $Rxy \subset Rx \subset R$ and $Rx/Rxy \approx R/Ry$, we see that $|xy| = |x||y|$. We will also need (3) $|x + y| \leq |x| + |y|$, (4) there is an $\alpha > 0$ such that for all positive integers n, the number of $x \in R$ with $|x| \leq n$ is $\geq \alpha n$, and (5) for any $n < \infty$ there are only a finite number of ideals (x) with $|x| \leq n$.

For R = Z, $|x|$ is the usual absolute value and all these properties are clear. For R = k[X], with $k = F_q$, we have $|f(X)| = q^{\deg f}$. Therefore $|x + y| \leq \max(|x|,|y|)$. In (4), $|f| \leq n \Rightarrow \deg f \leq \log_q n$. If $\chi = [\log_q n]$, there are $q^{\chi+1}$ such f. But $\chi + 1 \geq \log_q n$ so we may take $\alpha = 1$. (5) is clear.

The theorem is proved in 4 steps, the final three of which use only (1) and (2). It would be interesting to know to what extent (3), (4) and (5) are really needed.

Let K be the quotient field of R and let A be a finite dimensional, semisimple K-algebra. Let Λ be an order of A over R. We want to show that there are only a finite number of isomorphism classes of Λ-modules M which are finitely generated and torsion free over R of rank $\leq$ s. We will reduce the problem to the case where A is simple, then to the case where A is a division ring, and then to the case $KM \approx A$. We then apply the classical finiteness of class number argument from algebraic number theory.

Lemma 3.11. If Λ, Γ are R-orders in A, the theorem is true for Λ if and only if it is true for Γ.

Proof. We first observe that $\Lambda \cap \Gamma$ is an order. It is clearly an R-algebra and finitely generated over R since R is noetherian. If $a \in A$ we can find r, $s \in R$, r, $s \neq 0$ so $ra \in \Lambda$, $sa \in \Gamma$. Now $rsa \in \Gamma \cap \Lambda$ so we have $K(\Lambda \cap \Gamma) = A$.

Therefore it will suffice to consider the case $\Gamma \subset \Lambda$. Suppose the theorem holds for Γ. Let $M_1, \ldots, M_r$ represent all isomorphism classes of Γ-modules finitely generated and torsion free of rank $\leq s$ over R. If M is a Λ-module with the same properties, there will be an isomorphism $f: M \approx M_i$ over Γ. This extends uniquely to an A-isomorphism $f': KM \approx KM_i$ and $f'|\Lambda M: \Lambda M \approx \Lambda M_i$. But $\Lambda M = M$ and so $\Lambda M_i = M_i$ in KM_i. The action of Λ on this is uniquely determined by the action of Γ on M_i. Therefore there are only a finite number of possible isomorphism classes for M.

Now assume the theorem holds for Λ. Let $N_1, \ldots, N_r$ represent the isomorphism classes of Λ-modules, finitely generated and torsion free of rank $\leq s$ over R. If M is a Γ-module with the same properties, form $\Lambda M \subset KM$. This is a Λ module with the specified properties so $\Lambda M \approx N_i$ for some i. Since Λ is finitely generated as an R-module and $K\Lambda = K\Gamma$, there is some $t \neq 0$ in R with $t\Lambda \subset \Gamma$. Therefore $t\Lambda M \subset M \subset \Lambda M$ since $\Gamma M = M$. By (1) ΛM is a free R-module of rank $m \leq s$. Therefore

$\Lambda M/t\Lambda M \approx \coprod_1^m R/Rt$ which is finite by (2). If ΛM is known, this allows only a finite number of possibilities for M. Since there are also only a finite number of possibilities for ΛM, we are done.

Now let $A = A_1 \times \ldots \times A_n$. If Λ_i is an order in A_i, then $\Lambda = \Lambda_1 \times \ldots \times \Lambda_n$ is an order in A. Any Λ module M has the form $M = M_1 \times \ldots \times M_n$ where M_i is a Λ_i-module ($M_i = e_i M$ where $e_i = (0, \ldots, 0, 1, 0 \quad 0)$ with a 1 in the i-th place). If the

Jordan-Zassenhaus theorem holds for all Λ_i it will clearly hold for Λ. Therefore we can assume A is simple.

Let $A = M_n(D)$ where D is a division algebra. Let Γ be an order in D. By Lemma 3.11 again, it will suffice to consider $\Lambda = M_n(\Gamma)$. The Morita theorems give a one to one correspondence between Γ-modules and Λ modules. This may be done very explicitly here. Let ${}_\Gamma P_\Lambda = \{(x_1, \dots, x_n) \mid x_i \in \Lambda\}$ and ${}_\Lambda Q_\Gamma = \{\begin{pmatrix} y_1 \\ \vdots \\ y_n \end{pmatrix} \mid y_i \in \Gamma\}$. Define $\gamma : P \otimes_\Lambda Q \to \Gamma$ by $(x_i) \otimes (y_j) \mapsto \sum x_i y_i$ and $\lambda : Q \otimes_\Gamma P \to \Lambda$ by $(y_i) \otimes (x_j) \mapsto (y_i x_j)$. These are clearly Γ, Γ resp Λ, Λ - bimodule homomorphisms. They are actually isomorphisms. To see this for λ note that $P \approx \coprod_1^n \Gamma$, $Q \approx \coprod_1^n \Gamma$ as Γ modules. Therefore $Q \otimes_\Gamma P \approx \coprod_1^n \coprod_1^n \Gamma \otimes_\Gamma \Gamma = \coprod_1^n \coprod_1^n \Gamma$. This is obviously isomorphic to Λ and we easily check that λ is the obvious isomorphism. For γ note that P is generated as a Λ module by (1, 0, ..., 0). The annihilator of this is the right ideal I of all matrices with first row 0. Therefore $P \approx \Lambda/I$ so $P \otimes_\Lambda Q \approx Q/IQ$. Now IQ consists of all $(y_i) \in Q$ with $y_1 = 0$. Therefore $Q/IQ \approx \Gamma$ and we can check that this isomorphism is given by γ. Now let ${}_\Gamma M$, ${}_\Lambda M$ be the categories of left Γ and Λ modules and define $S: {}_\Gamma M \to {}_\Lambda M$ by $S(M) = Q \otimes_\Gamma M$, and $T: {}_\Lambda M \to {}_\Gamma M$ by $T(N) = P \otimes_\Lambda N$. Then $TSM \approx M$

and STN $\approx$ N by γ and λ so S and T are inverse equivalences. This argument works for any ring Γ. For the R-order considered above, we observe that $S(M) = Q \otimes_{\Gamma} M \approx \coprod_{1}^{n} M$ as an R-module. Therefore M is finitely generated and torsion free if and only if S(M) is. Also a bound on the rank of one implies a bound on the rank of the other. Therefore, if the Jordan-Zassenhaus theorem holds for Γ, it also holds for Λ.

We can now assume that A is a division algebra. We next reduce to the case of Λ-modules M with KM $\approx$ A. If M is any Λ module finitely generated and torsion free over R of rank s, the A module KM has a composition series $KM = W_0 \supset W_1 \supset \ldots \supset W_s = 0$. Let $M_i = W_i \cap M$. Then $M = M_0 \supset M_1 \supset \ldots \supset M_s = 0$ and $M_i/M_{i+1} \subset K(M_i/M_{i+1}) = W_i/W_{i+1} \approx A$ since there is only one simple A-module. If the Jordan-Zassenhaus theorem holds for s = 1, there will be only a finite number of possibilities for each M_i/M_{i+1}. By induction on s we can assume there are only a finite number of possibilities for M_1 and we must show that there are only a finite number of possible extensions $0 \to M_1 \to M \to M/M_1 \to 0$.

Lemma 3.12. Let R satisfy (1) and (2). Let A be a semisimple K-algebra and Λ an R order in A. Let X and Y be finitely generated Λ-modules, torsion free over R. Then there are only a finite number of possible extensions $0 \to Y \to E \to X \to 0$.

Proof. We recall that such extensions are classified by the elements of $\mathrm{Ext}^1_{\Lambda}(X, Y)$. This is done explicitly as follows. Let

$0 \to B \xrightarrow{i} P \xrightarrow{j} X \to 0$ with P finitely generated and projective over Λ. We can define $\mathrm{Ext}^1_\Lambda(X, Y)$ as the cokernel of Λ*
$\mathrm{Hom}_\Lambda(P, Y) \to \mathrm{Hom}_\Lambda(B, Y)$

$$(*) \qquad \mathrm{Hom}_\Lambda(P, Y) \xrightarrow{i^*} \mathrm{Hom}_\Lambda(B, Y) \xrightarrow{\delta} \mathrm{Ext}^1_\Lambda(X, Y) \to 0$$

Given an extension $0 \to Y \xrightarrow{p} E \xrightarrow{q} X \to 0$, the projectivity of p shows that there is a map Θ making the diagram

$$(*) \qquad \begin{array}{ccccccccc} 0 & \to & B & \xrightarrow{i} & P & \xrightarrow{j} & X & \to & 0 \\ & & \downarrow{\scriptstyle\varphi} & & \downarrow{\scriptstyle\Theta} & & \| & & \\ 0 & \to & Y & \xrightarrow{p} & E & \xrightarrow{q} & X & \to & 0 \end{array}$$

commute. Any two such Θ's will differ by a map p^Ψ where $\Psi : P \to Y$. The corresponding φ's differ by $\Psi i = i^*(\Psi)$ so $\xi = \delta(\varphi) \in \mathrm{Ext}^1_\Lambda(X, Y)$ is well defined. Conversely, given $\xi \in \mathrm{Ext}^1_\Lambda(X, Y)$ define the extension $0 \to Y \to E \to X \to 0$ by forming the pushout of the diagram

$$\begin{array}{ccccccccc} 0 & \to & B & \xrightarrow{i} & P & \xrightarrow{j} & X & \to & 0 \\ & & \downarrow{\scriptstyle\varphi} & & & & & & \\ & & Y & & & & & & \end{array}$$

where φ is any element of $\mathrm{Hom}_\Lambda(B, Y)$ with $\delta(\varphi) = \xi$. This extension clearly has class ξ. If we have two extensions with the class ξ we can assume the map φ is the same for both by replacing Θ by $\Theta + p\Psi$ if necessary for one of the extensions. Since * is easily seen to be a pushout diagram, the extension is determined by φ and so by ξ.

In the case of interest, B and Y are finitely generated over R and therefore so is $\mathrm{Ext}^1_\Lambda(X, Y)$. If we can show that this is a torsion module, it will be finite by (2) since (1) implies it is a sum of cyclic modules.

Tensoring (*) with K and using the fact that everything is finitely generated over R, we get

$$\mathrm{Hom}_A(KP, KY) \to \mathrm{Hom}_A(KB, KY) \to K \otimes_R \mathrm{Ext}^1_\Lambda(X, Y) \to 0 .$$

Therefore $K \otimes_R \mathrm{Ext}^1_\Lambda(X, Y) \approx \mathrm{Ext}^1_A(KX, KY)$ but this is 0 since all extensions split over A which is semisimple. Therefore $\mathrm{Ext}^1_\Lambda(X, Y)$ is torsion.

We are now reduced to the case where A is a division algebra and $KM \approx A$. We may assume $M \subset A$. Since M is finitely generated, there is some $t \neq 0$ in R such that $I = tM \subset \Lambda$. Thus we must show there are only a finite number of isomorphism classes of left ideals I in Λ.

By (1), Λ is a free R-module. Let $w_1, \ldots, w_n$ be a base for it. If $x \in \Lambda$, use (1) to reduce the matrix of $\Lambda \xrightarrow{\hat{x}} \Lambda$ (by $y \mapsto yx$) to diagonal form. This shows that $\Lambda/\Lambda x$ as an R-module is a direct sum of modules R/Rd, and $\prod d_i = \det \hat{x}$. Since the order of R/Rx_i is $|x_j|$ by definition, the order of $\Lambda/\Lambda x$ is $\prod |d_i| = |\det \hat{x}|$. This is non 0 for $x \neq 0$ since A is a division algebra and so $A \xrightarrow{\hat{x}} A$ is an isomorphism.

Write $x = x_1w_1 + \ldots + x_nw_n$ with $x_i \in R$. If $w_iw_j = \sum a_{ijk}w_k$, then $w_ix = \sum_{j,k} x_ja_{ijk}w_k$. Therefore det $\hat{x}$ will be a homogeneous polynomial $f(x_1, \ldots, x_n)$ of degree n in $x_1, \ldots, x_n$. Using (3), we see that there is a constant C such that $|x_i| \leq N$ for all i implies $\#(\Lambda/\Lambda x) = |\det \hat{x}| = |f(x_1, \ldots, x_n)| \leq CN^n$.

Now $[\Lambda: I] = \#\Lambda/I$ is finite by (1) and (2) since it is a torsion module and finitely generated. Choose an integer k so that $k^n \leq [\Lambda:I] < (k+1)^n$. Let α be as in (4) and choose an integer χ with $k/\alpha + 1 \geq \chi > k/\alpha$. Let $S = \{x \in R \mid |x| \leq \chi\}$. By (4), $\#S \geq \alpha\chi > k$ so $\#S \geq k + 1$. Therefore the number of $x = x_1w_1 + \ldots + x_nw_n$ with all $x_i \in S$ is $\geq (k+1)^n > [\Lambda: I]$ so there are two such x's congruent mod I. Subtracting them gives $z \in I$, $z \neq 0$, $z = z_1w_1 + \ldots + z_nw_n$ where each z_i has the form $s - s'$ with $s, s' \in S$. By (3), $|z_i| \leq 2\chi$. Therefore $[\Lambda: \Lambda z] \leq C(2\chi)^n$. Since $\Lambda \supset I \supset \Lambda z$, we have $[I: \Lambda z] = [\Lambda: \Lambda z][\Lambda: I]^{-1} \leq C(2\chi)^n k^{-n} = C(\frac{2\chi}{k})^n$ but $\chi \leq 1 + k\alpha^{-1}$ so $\chi/k \leq 1/k + \alpha^{-1} \leq 1 + \alpha^{-1}$. Therefore $[I: \Lambda z] \leq D$ where $D = C(2 + 2\alpha^{-1})^n$ is independent of I. We need one more lemma in order to apply this.

<u>Lemma 3.13</u>. Let R satisfy (1), (2), (4) and (5). Given $D < \infty$, there is an $r \in R$, $r \neq 0$ such that r annihilates every R-module C with $\#(C) \leq D$.

Proof. By (1), $C \approx \coprod R/Rd_i$. Since (4) implies that $\#R = \infty$, all $d_i \neq 0$. Now $d = \prod d_i$ annihilates C and $|d| = \prod |d_i| = \#C \leq D$. By (5) there are a finite number of elements $b_1, \ldots, b_k \in R$ such that every d with $|d| \leq D$ has the form $d = \varepsilon b_i$, ε a unit of R. Clearly $r = b, b_2 \ldots b_k$ will do.

Now if I is any left ideal of Λ, we have found a $z \in \Lambda$ so $[I: \Lambda z] \leq D$. Therefore $[Iz^{-1}: \Lambda] = [I: \Lambda z] \leq D$. Let r be as in Lemma 3.13. Then $r(Iz^{-1}/\Lambda) = 0$ so $I \approx J = rIz^{-1}$ and $\Lambda \supset J \supset \Lambda r$. But $\Lambda/\Lambda r$ is finite of order $|r|n$ where n is the rank of Λ. This allows only a finite number of possibilities for J.

Example. Theorem 3.9 is false if A is not semisimple. Let $L = Z[x]/(x^2)$, $A = Q[x]/(x^2)$, and $M_n = Z \oplus Z$ where x acts by the matrix $\begin{pmatrix} 0 & n \\ 0 & 0 \end{pmatrix}$. These are all distinct for $n \geq 0$. We only need to show that we can recover n from M_n. Let u and v be the two basic vectors. Then $xu = nv$ and $xv = 0$. Hence $\ker(\hat{x})$ $\hat{x}: M_n \to M_n$ is Zv and the image $(\hat{x}) = Znv$. Therefore, $\ker \hat{x}/\text{im } \hat{x} \cong Z/nZ$ and we can recover n from M_n. Note that $QM_n \approx A$ for all n by $u \mapsto 1$, $v \mapsto \frac{1}{n}x$.

Chapter Four: Results on K_0 and G_0

Theorem 4.1. Let R be a Dedekind ring for which the Jordan-Zassenhaus theorem holds, K the quotient field of R, and π a finite group of order prime to char K. Then the kernel of $G_0(R\pi) \to G_0(K\pi)$ is finite.

<u>Proof</u>. (Rim) We have an exact sequence

$$0 \to C_0(R\pi) \to K_0(R\pi) \to K_0(K\pi)$$

where $C_0(R\pi)$ is defined to be the kernel. $C_0(R\pi)$ is finite by Theorem 3.8. Let X be the kernel of $G_0(R\pi) \to G_0(K\pi)$. We get a diagram

$$\begin{array}{ccccccccc} 0 & \to & C_0(R\pi) & \to & K_0(R\pi) & \to & K_0(K\pi) & & \\ & & \downarrow & & \downarrow & & \downarrow & & \\ 0 & \to & X & \to & G_0(R\pi) & \to & G_0(K\pi) & \to & 0 \end{array}$$

where the right hand map is an isomorphism. Since $G_0(R\pi)$ is a finitely generated abelian group, X will be finite if it is torsion. Hence it is enough to show its generators have finite order. We have an exact sequence

$$\coprod_{\substack{p \subset R \\ p \neq 0}} G_0(R/p\pi) \to G_0(R\pi) \to G_0(K\pi) \to 0$$

where the left hand map has image X. The map $\gamma\colon K_0(R/p\pi) \to G_0(R/p\pi)$ has finite coker by Theorem 2.20, say of order s. Pick $[M] \in G_0(R/p\pi)$. $s[M]$ lifts to $K_0(R/p\pi)$ and $[M]$ will have finite order if and only if $s[M]$ does. Therefore, it is enough to show that the image of $K_0(R/p\pi) \to G_0(R\pi)$ is torsion. Let M be a projective $R/p\pi$ module. We claim $\operatorname{pd}_{R\pi} M \leq 1$. M is a direct summand of a free $R/p\pi$ module. Hence, it is enough to show

$\mathrm{pd}_{R\Pi} R/p\Pi \leq 1$. $R/p\Pi = R/p \otimes_R R\Pi$. The sequence $0 \to p \to R \to R/p \to 0$ is exact. Tensoring with $R\Pi$ over R preserves exactness since $R\Pi$ is a free R module. Hence $0 \to p \otimes_R R\Pi \to R \otimes_R R\Pi \to R/p \otimes_R R\Pi \to 0$ is exact. But $p \otimes_R R\Pi$ and $R \otimes_R R\Pi$ are projective $R\Pi$ modules since p and R are projective R modules. Hence $\mathrm{pd}_{R\Pi} R/p\Pi \leq 1$ as claimed. Next we claim that if M is a finitely generated torsion module over R with $\mathrm{pd}_{R\Pi} M \leq 1$ then [M] lies in the image of $C_0(R\Pi) \to G_0(R\Pi)$. Since $C_0(R\Pi)$ is finite, this will show that [M] has finite order in $G_0(R\Pi)$ and will complete the proof. Since $\mathrm{pd}_{R\Pi} M \leq 1$ we can find an exact sequence $0 \to P \to F \to M \to 0$ of $R\Pi$ module with F finitely generated free and P finitely generated projective. Then $[M] = [F] - [P]$ lifts to $K_0(R\Pi)$. If $[F] - [P]$ goes to 0 in $K_0(K\Pi)$ then $[F] - [P] \in C_0(R\Pi)$ as desired. Tensoring over R with K which is flat over R we get

$$0 \to K \otimes_R P \to K \otimes_R F \to K \otimes_R M \to 0 \quad \text{exact.}$$

But $K \otimes_R M = 0$ since M is torsion and, hence, $[F] - [P] = 0$ in $K_0(K\Pi)$ as desired.

As an application of this we give a generalization of the Herbrand quotient. Suppose Π has periodic cohomology of period q, e.g., Π cyclic and $q = 2$. If M is a finitely generated $Z\Pi$ module, let $h^i(\Pi, M)$ be the order of $H^i(\Pi, M)$. Let $Q(M) = \prod_{i=1}^{q} h^i(\Pi, M)^{(-1)^i}$. If $0 \to M' \to M \to M'' \to 0$, is exact, then the exact cohomology sequence shows that $Q(M) = Q(M')Q(M'')$. Therefore Q defines a map $Q\colon G_0(Z\Pi) \to Q^{*+}$, the multiplicative group of rational numbers > 0. Since Q^{*+} is torsion free and $X = \ker[G_0(Z\Pi) \to G_0(K\Pi)]$ is finite, $Q(X) = 0$. Therefore Q factors through a map $Q\colon G_0(K\Pi) \to Q^{*+}$. If M is finite, $Q \otimes M = 0$ so $Q(M) = Q([K \otimes M]) = Q(0) = 1$. If Π is cyclic, $q = 2$, $Q(M)$ is the Herbrand quotient $h_{2/1}(M)$ and we have reproved Herbrand's result $h_{2/1}(M) = 1$ for finite M.

__Theorem 4.2.__ Let R be a Dedekind ring of characteristic 0 with quotient field K and let Π be a finite group of order n. If no prime dividing n is a unit in R and P is a finitely generated projective $R\Pi$ module, then $K \otimes_R P$ is free over $K\Pi$.

__Proof.__ We will prove this by computing characters. Let $F = \coprod_1^s K\Pi$. Then $\chi_F = s\chi_{K\Pi}$. If $g \in \Pi$ then $\hat{g}\colon K\Pi \to K\Pi$ given by $\hat{g}(t) = gt$ has trace equal to the number of $t \in \Pi$ such that $gt = t$. So

$$\chi_{K\Pi}(g) = \begin{cases} 0 & \text{if } g \neq 1 \\ n & \text{if } g = 1 \end{cases}. \quad \text{Hence } \chi_F(g) = \begin{cases} 0 & \text{if } g \neq 1 \\ sn & \text{if } g = 1 \end{cases}.$$

We need to show that the character of $K \otimes_R P$ is equal to χ_F for some F. Since $\chi_P(1) = \dim_K(P)$ we will need to show that n divides the rank of P.

__Lemma 4.3.__ If k is a field of characteristic p, Π is a finite p group, and I is the kernel of the augmentation $\varepsilon\colon k\Pi \to k$, then I is nilpotent.

__Proof.__ If G is a group and N a normal subgroup let J be the kernel of the map $RG \to RG/N$. As a left ideal J is generated by all $n - 1$

with $n \in N$. As an R module J is generated by all $xn - x$ with $x \in G$ and $n \in N$. Since Π is a p group there exists $1 \neq x \in \Pi$ which is central and of order p. Let $N = \langle x \rangle$. We have a commutative diagram with exact row

$$\begin{array}{ccccccccc} 0 & \rightarrow & J & \rightarrow & k\Pi & \rightarrow & k\Pi/\langle x\rangle & \rightarrow & 0 \\ & & & & {\scriptstyle \varepsilon}\searrow & & \swarrow & & \\ & & & & & k & & & \end{array}$$

which shows $J \subset I_{\Pi}$. We use induction on the number of elements in Π. Hence $(I_{\Pi}/J)^m = 0$ or equivalently $I_{\Pi}^m \subset J$. Therefore it is enough to show that J is nilpotent. But J is generated by $x^n - 1 = (x - 1)$ (something) for all n. Thus, $J = k\Pi(x - 1)$. Therefore, $J^n = k\Pi(x - 1)k\Pi(x - 1)\ldots$ but $(x - 1)$ is central. Hence $J^n = k\Pi(x - 1)^n$. So $J^p = k\Pi(x - 1)^p = k\Pi(x^p - 1) = 0$ since k has characteristic p. This completes the proof.

Corollary 4.4. Under the above assumptions $k\Pi$ is local.

Corollary 4.5. Under the above assumptions all finitely generated projective $k\Pi$ modules are free.

Proposition 4.5. (Nakayama) Let R be a Dedekind domain of characteristic 0, Π a finite group of order n, and P a finitely generated projective $R\Pi$ module. If no prime dividing n is a unit in R then n divides $\mathrm{rank}_R P$.

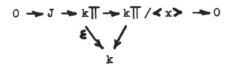

Proof. Let $p \mid n$ and Π_p be a Sylow p subgroup of Π. Then P is a finitely generated projective $R\Pi_p$ module. Hence if the theorem is true for p groups it is true for finite groups. Let $p \mid n = p^a$ then p is not a unit. Hence there is a prime ideal $\mathfrak{p} \subset R$ such that $p \in \mathfrak{p}$. $P/\mathfrak{p}P = R/\mathfrak{p} \otimes_R P$ is a projective $R/\mathfrak{p}\Pi$ module. Therefore, $P/\mathfrak{p}P$ is free over $R/\mathfrak{p}\Pi$ by corollary 4.5 so $|\Pi| \mid \dim_{R/\mathfrak{p}} P/\mathfrak{p}P$. As an R module $P = A_1 \oplus \dots \oplus A_s$ where A_i are ideals. Then $P/\mathfrak{p}P = A_1/\mathfrak{p}A_1 \oplus \dots \oplus A_s/\mathfrak{p}A_s = R/\mathfrak{p} \oplus \dots \oplus R/\mathfrak{p}$. Therefore, $\dim_{R/\mathfrak{p}} P/\mathfrak{p}P = \operatorname{rank}_R P$. This completes the proof of proposition 4.6.

Definition. Let R be a ring, Π a finite group, and M be a $R\Pi$ module. Then $\underline{M^{\Pi}} = \{x \in M \mid sx = x \text{ for all } s \in \Pi\}$ and $\underline{M/\Pi} = M/IM = M/((s-1)M)_{s \in \Pi}$. That is, M/Π is obtained from M by identifying x and sx for all $s \in \Pi$.

Let $N \in R\Pi$ be the element $N = \sum_{s \in \Pi} s$. If $t \in \Pi$, $tN = N = Nt$. If $x \in M$, then $Nx = \sum sx$ and $tNx = Nx$. Thus $Nx \in M^{\Pi}$ for all $x \in M$. Therefore, $\hat{N}: M \to M^{\Pi} \subset M$. $N(s-1)x = (Ns - N)x = 0$. Therefore, $\hat{N}$ factors

$$\begin{array}{ccc} M & \xrightarrow{N} & M^{\Pi} \\ & \searrow \quad \nearrow & \\ & M/\Pi & \end{array}$$

Lemma 4.7. If M is projective then the induced map $\hat{N}: M/\Pi \to M^{\Pi}$ is an isomorphism.

Proof. There exists M' such that $M \oplus M' = F$ is free. Clearly $M^{\Pi} \oplus M'^{\Pi} = F^{\Pi}$ and $M/\Pi \oplus M'/\Pi = F/\Pi$ and if $\hat{N}: F/\Pi \to F^{\Pi}$ is an isomorphism then $\hat{N}: M/\Pi \to M^{\Pi}$ is also. $F^{\Pi} = \coprod (R\Pi)^{\Pi}$ and $F/\Pi = \coprod (R\Pi/\Pi)$. Hence we only need to prove the lemma for $M = R\Pi$. $R\Pi/\Pi = R\Pi = R\Pi/I = R$ where the map $R\Pi/\Pi \to R$ is given by the augmentation. If $\sum a_s s \in (R\Pi)^{\Pi}$. Then $t\sum a_s s = \sum a_s s$. Therefore, $\sum a_s ts = \sum a_{t^{-1}p}\, p = \sum a_{t^{-1}s}\, s$ where $p = ts$. Or $a_s = a_{t^{-1}s}$ for all s,t. That is, $a_s = a$ for all $s \in \Pi$. Hence, $(R\Pi)^{\Pi} = R.N$. Now the map $\hat{N}: R\Pi/\Pi \to (R\Pi)^{\Pi}$ clearly sends $1 \in R\Pi/\Pi = R$ to $N \in (R\Pi)^{\Pi}$ and is an isomorphism.

Lemma 4.8. If Π is a finite solvable group of order n, R is a Dedekind ring of characteristic 0 such that no prime dividing n is a unit in R, and P is a finitely generated projective $R\Pi$ module, then $\text{rank}_R P = n\ \text{rank}_R P^{\Pi}$.

Proof. Lemma 4.7 allows us to use P/Π for P^{Π}. We proceed by induction on $|\Pi|$.

1) If Π is not of prime order, there is a proper normal subgroup N of Π. $\text{rank}_R P = |N| \text{rank}_R(P/N)$ by inductive hypothesis.

$P/N = R(\Pi/N) \otimes_{R\Pi} P = P/I_N P = R \otimes_{RN} P = R \otimes_{RN} (R\Pi \otimes_{R\Pi} P) = (R \otimes_{RN} R\Pi) \otimes_{R\Pi} P = R(\Pi/N) \otimes_{R\Pi} P$. The last step is true because $R \otimes_{RN} R\Pi = R\Pi/(n-1)R\Pi$ for all $n \in N$. Therefore, $R \otimes_{RN} (R\Pi \otimes_{R\Pi} P) = R(\Pi/N) \otimes_{R\Pi} P$ is projective over $R\Pi/N$,

$$\begin{aligned} \text{rank}_R P/N &= |\Pi/N|\, \text{rank}_R(P/N/\Pi/N) \\ &= |\Pi/N|\, \text{rank}_R(P/\Pi) \end{aligned}$$

by the inductive hypothesis. Hence $\mathrm{rank}_R P = |\Pi|\, \mathrm{rank}_R P^{\Pi}$ for $|\Pi|$ not prime.

2) Π is cyclic of prime order p. Then there is a prime ideal $y \subset R$ with $p \in y$. We reduce mod y. Then $(P/\Pi)/y(P/\Pi) = (P/yP)/\Pi$ since change of rings is transitive. Now $(P/yP)/\Pi$ is a projective $R/y\Pi$ module. By corollary 4.5, $(P/yP)/\Pi$ is a free $R/y\Pi$ module. Therefore, $\mathrm{rank}_R P/\Pi = \dim_{R/y}(P/\Pi)/y(P/\Pi) = \dim_{R/y}(P/yP)/\Pi$ while $\mathrm{rank}_R P = \dim_{R/y}(P/yP)$. Let $k = R/y$ and

$P/yP = \coprod_1^r k\Pi$. This has dim pr and $(P/yP)/\Pi = \coprod_1^r k\Pi/\Pi = \coprod_1^r k$ has dimension r. This completes the proof of lemma 4.8. Now we return to the proof of theorem 4.2. R is a characteristic 0 Dedekind ring, Π a finite group of order n, and no prime dividing n is a unit in R. P is a finitely generated projective $R\Pi$ module. We know that $\mathrm{rank}_R P = ns$ for some integer s. Let F be a free $R\Pi$ module on s generators. Then $\mathrm{rank}_R F = \mathrm{rank}_R P$. We need to show $K \otimes_R F$ is isomorphic over $K\Pi$ to $K \otimes_R F$. It is enough to show they have the same characters. Each element of Π generates a cyclic subgroup of Π. Thus if $K \otimes_R P$ is isomorphic to $K \otimes_R F$ as a $K\Pi'$ module for every cyclic $\Pi' \subset \Pi$, then they have the same characters. Thus, we can assume Π is cyclic. Now $K\Pi$ is semisimple. Let $S_1, \ldots, S_r$ be representatives of all the isomorphism classes of simple modules,

$K \otimes_R P = \coprod_1^r \coprod_1^{n_i} S_i$ and $K \otimes_R F = \coprod_1^r \coprod_1^{m_i} S_i$. We only need to show $n_i = m_i$ for all i.

By Schur's lemma, if S and T are non isomorphic $K\Pi$ modules, then $\mathrm{Hom}_{K\Pi}(S, T) = 0$ while $\mathrm{Hom}_{K\Pi}(S, S) \neq 0$. If $M = \coprod_1^r \coprod_1^{n_i} S_i$, then $\mathrm{Hom}_{K\Pi}(S_i, M) = \coprod_1^{n_i} \mathrm{Hom}_{K\Pi}(S_i, S_i)$. Hence,

$$n_i = \frac{\dim_K \mathrm{Hom}_{K\Pi}(S_i, M)}{\dim_K \mathrm{Hom}_{K\Pi}(S_i, S_i)}.$$

Thus M is $K\Pi$ isomorphic to N if and only if $\mathrm{Hom}_{K\Pi}(S_i, M)$ and $\mathrm{Hom}_{K\Pi}(S_i, N)$ have the same dimension over K for all i. Thus it is enough to show that $\dim_K \mathrm{Hom}_{K\Pi}(S_i, K \otimes_R P)$ depends only on $\mathrm{rank}_R P$ for all i.

We define a Π action on $\mathrm{Hom}_K(S, M)$ by $g(f)(s) = gf(g^{-1}s)$ for all $g \in \Pi$, $f \in \mathrm{Hom}_K(S, M)$. It is clear that $\mathrm{Hom}_{K\Pi}(S, M) = \mathrm{Hom}_K(S, M)^{\Pi}$ with respect to this action. Let $S^* = \mathrm{Hom}_K(S, K)$. Then $S^* \otimes M$ is isomorphic to $\mathrm{Hom}_K(S, M)$ by the map sending $f \otimes m$ to the function g with $g(s) = f(s)m$. This is a $K\Pi$ isomorphism. Let $p \in \Pi$. Then $p(f \otimes m) = pf \otimes pm$ and $pf \otimes pm \rightsquigarrow h: S \to N$ is given by $h(s) = p(f)(s)pm = pf(p^{-1}s)pm = f(p^{-1}s)pm$ since Π acts trivially on K. Next we compute the action of p on g = image of $f \otimes m$. $p(g)(s) = pg(p^{-1}s) = p(f(p^{-1}s)m) = f(p^{-1}s)pm$. Hence the map is a Π map.

Thus, $\mathrm{Hom}_{K\Pi}(S, M) = (S^* \otimes M)^{\Pi}$ and by lemma 4.7 $\mathrm{Hom}_{K\Pi}(S, M)$

is isomorphic to $(S^* \otimes M)/\Pi$. S^* is a finitely generated projective $K\Pi$ module, therefore there exists a finitely generated $R\Pi$ module $T \subset S^*$ such that $K \otimes T$ is isomorphic to S^*. We need only let T be the $R\Pi$ module generated by a finite set of generators for S^* over $K\Pi$. To compute the rank of $\mathrm{Hom}_{K\Pi}(S, K \otimes_R P)$ we note that

$(S^* \otimes K \otimes_R P)/\Pi = (K \otimes T \otimes K \otimes P)/\Pi = (K \otimes T \otimes P)/\Pi$.

$\dim_K K \otimes (T \otimes P)/\Pi = \mathrm{rank}(T \otimes P)/\Pi$. Since P is projective over $R\Pi$, so is $T \otimes P$. Therefore $\mathrm{rank}(T \otimes P)/\Pi = \frac{1}{n}\,\mathrm{rank}(T \otimes P) = \frac{1}{n}$ rank T rank P. Hence we can compute the $\dim_K \mathrm{Hom}_{K\Pi}(S_i, P)$ solely from rank P. This completes the proof of theorem 4.2.

<u>Proposition 4.9</u>. Let R be a Dedekind ring with quotient field K, Π a finite group of order n and L an R order of $K\Pi$. Then $nL \subset R\Pi$.

<u>Remark</u>. If characteristic of K does not divide n, then $L \subset \frac{1}{n} R\Pi$. Since $R\Pi$ is a noetherian R module, this gives an easy proof that there exist maximal orders in $K\Pi$.

<u>Proof</u>. We examine the trace map tr: $K\Pi \to K$ given by $\mathrm{tr}(x) = \mathrm{tr}(\hat{x})$ where $\hat{x}(a) = xa$. We need some information on the image of tr.

<u>Lemma 4.10</u>. If L is an R order, then tr: $L \to R$.

<u>Proof</u>. It is enough to check this locally since $\bigcap R_p = R$. Then L_p is a free R_p module. Let $\{w_i\}$ be a basis for L_p. Then $xw_i \in L_p$. $\{w_i\}$ is also a basis for $K\Pi$ over K. To compute trace we write $xw_i = \sum a_{ij}w_j$ with $a_{ij} \in K$. Then $\mathrm{tr}(\hat{x}) = \sum a_{ii}$. $xw_i \in L_p$ implies $a_{ii} \in R_p$. Thus $\mathrm{tr}(\hat{x}) \in R_p$ as desired.

To complete the proof we compute trace using the $s \in \Pi$ as a basis. The coefficient of s in ts is 0 if $t \neq 1$ for $t \in \Pi$ and coefficient of s in 1s is 1. Thus $tr(t) = 0$ if $t \neq 1$ and $tr(1) = n$. Trace is additive, therefore $tr(\sum_{s \in \Pi} a_s s) = na_1$. Given $k \in L$ we want to show $nk \in R\Pi$. Let $k = \sum a_s s$ the $na_1 = tr(k) \in R$. If $t \in \Pi$, then $t^{-1}k \in L$, $t^{-1}k = \sum a_s t^{-1}s$, and so $tr(t^{-1}k) = na_t \in R$ for every $t \in \Pi$. Thus, $nk = \sum na_t t \in R\Pi$ as desired.

Let Π be a finite abelian group of order n and R a Dedekind ring with quotient field K whose characteristic does not divide n. Pick $x \in K\Pi$ which is integral over $R\Pi$. Then $R\Pi[x]$ is a finitely generated $R\Pi$ (and hence also R) module. Therefore, $R\Pi[x] = \Lambda$ is an R order. By the previous proposition, $\Lambda \subset \frac{1}{n}R\Pi$. Let Γ be the integral closure of $R\Pi$ in $K\Pi$. Then, by the above, $\Gamma \subset \frac{1}{n}R\Pi$ and hence is finitely generated over R. Thus, Γ is an R order of $K\Pi$. But any order has to be finitely generated and integral over $R\Pi$. Therefore, Γ is a maximal order and even the unique maximal order of $K\Pi$. Since $R\Pi$ is integral over R, Γ is also the integral closure of R in $K\Pi$.

$K\Pi = \coprod_{i=1}^{r} L_i$ where the L_i are fields. Let e_i be the unit of L_i. The e_i satisfy the equation $x^2 - x$ and hence are integral over R and, thus, in Γ. $1 = \sum e_i$ and $e_i e_j = 0$ if $i \neq j$ so

$\Gamma = \Gamma_1 x \ldots x \Gamma_r$ where $\Gamma_i = \Gamma e_i$. Since $\Gamma_i = L_i \cap \Gamma$, Γ_i is the integral closure of R in L_i. The Γ_i are all Dedekind rings. Now $K\Pi \to L_i$ is onto and L_i is generated over K by elements ζ_s, the images of the $s \in \Pi$ in L_i. $s^n = 1$ implies $\zeta_s^n = 1$. So L_i is generated over K by its n-th roots of 1. That is L_i is a cyclotemic extension of K. Now $\{x \in L_i | x^n = 1\}$ is a cyclic group which is generated by one root of 1 say ζ. Then $L = K(\zeta)$. If Π is cyclic of order n, then $K\Pi = K[x]/(x^n - 1)$. One of the fields L_i is $K(\zeta)$ where ζ is a primitive n-th root of 1. Our remarks thus yield the following result which does not involve group rings explicitly.

Corollary 4.11. Let R be a Dedekind ring, K its quotient field, and $L = K(\zeta)$ with ζ a primitive n-th root of 1. Let R' be the integral closure of R in L. Then $nR' \subset R[\zeta]$.

Remark. We might hope that $R' = R[\zeta]$. This is false in general.

Corollary 4.12. With the above notation, if p is a prime ideal of R that does not contain n, then p is unramified in R'.

Proof. Let $R'p = P_1^{e_1} \ldots P_r^{e_r}$. We want to show all the e_i are 1. Now $R'/R'p = \prod_1^r R'/P_i^{e_i}$. If we could show $R'/R'p$ is semisimple then we would be done because commutative semisimple rings have no nilpotent elements.

Nothing is changed by localizing at p. Hence we can assume n is a unit. Then, by corollary 4.11, $R' = R[\zeta]$ so

$R'/pR' = R[\zeta]/pR[\zeta]$. This is a quotient of $R/p\Pi$. Since the characteristic of R/p is prime to $|\Pi|$, $R/p\Pi$ is semisimple and thus so is R'/pR'.

Theorem 4.13. Let R be a Dedekind ring of characteristic 0, K its quotient field and $p \in z$ a prime such that

1) p is not a unit in R and

2) p is unramified in R (i.e., $(p) = P_1 \cdots P_r$ where the P_i all distinct).

Let ζ be a primitive p^n root of 1 (for any n), then

1) $[K(\zeta): K] = \varphi(p^n) = p^{n-1}(p - 1)$,

2) the integral closure R' of R in $K(\zeta)$ is $R[\zeta]$, and

3) $P_1, \ldots, P_r$ are totally ramified in R', i.e., $R'P_i = \mathcal{P}_i^{\varphi(p^n)}$.

Corollary 4.14. Let R be a Dedekind ring of characteristic 0, K its quotient field, $n \in Z$. If for all primes $p \in Z$ dividing n we have

1) p is not a unit in R and

2) p is unramified in R.

Then if ζ is a primitive nth root of 1, we have

1) $[K(\zeta): K] = \varphi(n)$,

2) the integral closure of R in $K(\zeta)$ is $R[\zeta]$, and

3) $p \subset R$ is ramified if and only if $n \in p$.

Proof of the corollary from the theorem. $n = p^v m$ where $p \nmid m$. $K \subset K(r) \subset K(\zeta)$ where r is a primitive mth root of 1. By induction on n we can go from K to K(r). To get from K(r) to $K(\zeta)$ we only need adjoin a primitive p^v-th root of 1. p is not a unit in R[r] since it is not in R and p does not ramify since $p \nmid m$. This proves the corollary.

Proof of Theorem 4.13. If $p \in P$, then $p \notin P^2$ since p is unramified in R. Hence $\text{ord}_P(p) = 1$ for all $P|p$. Let $R'P = \mathfrak{P}_1^{e_1} \dots \mathfrak{P}_r^{e_r}$ and $f_i = [R'/\mathfrak{P}_i : R/P]$. Then $\sum_{i=1}^{r} e_i f_i = [K(z): K]$ (since R has characteristic 0, there is no inseparability).

3) asserts that r = f = 1 or that

$$R'P = \mathfrak{P}^e \text{ where } e = [K(z): K].$$

Now
$$\Phi_{p^n}(x) = \prod_{\substack{\zeta \text{ primative} \\ p^n\text{-th root} \\ \text{of } 1}} (x - \zeta)$$

so
$$p = \Phi_{p^n}(1) = \prod_{\substack{\zeta \text{ primitive} \\ p^n\text{-th root} \\ \text{of } 1}} (1 - \zeta) \in R'.$$

Let z and $w = z^k$ be two primitive p^n-th roots of 1. Then $1 - w = (1 - z^k) = (1 - z)(1 + \dots + z^{k-1})$. Therefore, $1 - z | 1 - w$. Similarly $1 - w | 1 - z$. Thus $1 - w = (\text{unit})\ (1 - z)$. Since $p = \prod(1 - \zeta)$ as above, we have $p = (\text{unit})\ (1 - \zeta)^{\varphi(p^n)}$ for ζ a fixed p^n-th primitive root of 1. Let $\lambda = 1 - \zeta$. Then $(p) = (\lambda)^{\varphi(p^n)}$. Examining $\mathfrak{P}_1^{e_1}$ above we have for $x \in R$ we obtain $e_1 \cdot \text{ord}_P x = \text{ord}_{\mathfrak{P}_1} x$. Hence,

$e_1 \operatorname{ord}_p p = \operatorname{ord}_{\mathcal{P}_1} p = \varphi(p^n)\operatorname{ord}_{\mathcal{P}_1}(\lambda)$. But $\operatorname{ord}_p(p) = 1$ by assumption. So $e_1 = \varphi(p^n)\operatorname{ord}_{\mathcal{P}_1}\lambda$ and $e_1 \geq \varphi(p^n)$. But $\sum_{i=1}^{r} e_1 f_1 = [K(z): K] \leq \varphi(p^n)$. Thus $r = 1$, $e_1 = 1$, and $[K(z): K] = \varphi(p^n)$ and so $R'P = \mathcal{P}_1^{\varphi(p^n)}$ where $\mathcal{P}_1$ is the unique prime divisor of $R'P$. This proves (1) and (3). We also see that $\operatorname{ord}_{\mathcal{P}_i}\lambda = 1$ so $(\lambda) = \mathcal{P}_1 \ldots \mathcal{P}_r$ where $\mathcal{P}_i$ is as in (3).

Finally we want to show that $R' = R[z]$. By corollary 4.11 $p^n R' \subset R[z]$. Since we have seen that $f = 1$ for $\mathcal{P}_i$ over P, we have $R'/\mathcal{P}_i = R/P_i$ where $\mathcal{P}_i$ is the unique prime over P_i for $i = 1, \ldots, r$. If $x \in R'$, there exists $y_i \in R$ such that $x \equiv y_i \bmod \mathcal{P}_i$. By the Chinese Remainder Theorem on $P_1, \ldots, P_r$ in R there exists $y \in R$ such that $y \equiv y_i \bmod P_i$ $i = 1, \ldots, r$. Thus $y - x \in \bigcap_{i=1}^{r} \mathcal{P}_i = \prod_{i=1}^{r} \mathcal{P}_i = (\lambda)$. Consider $R'/R[\zeta]$ as a $R[\zeta]$ module. Since for any $x \in R'$, there is $y \in R$ with $x - y \in \lambda R'$ we know that

$$R'/R[\zeta] = \lambda(R'/R[\zeta]) = \ldots = \lambda^N(R/R[\zeta])$$

for all $N \geq 1$. But $\lambda^{\varphi(p^n)} = (\text{unit})p$ so $\lambda^{\varphi(p^n)t} = (\text{unit})p^t$. Thus $R'/R[\zeta] = p^t(R'/R[\zeta])$. Since $p^n R' \subset R[\zeta]$ and $p^n(R'/R[\zeta]) = R'/R[\zeta]$ we have $R'/R[\zeta] = 0$ and $R' = R[\zeta]$.

Corollary 4.15. Let R be a Dedekind ring of characteristic 0 and K its quotient field. Let $n \in Z$ such that for all primes $p|n$, p is not a unit in R and is unramified in R. Let ζ be a primitive n-th root of 1. Then $[K(\zeta):K] = \varphi(n)$ and the integral closure of R in $K(\zeta)$ is $R[\zeta]$.

<u>Proof</u>. Let $n = p_1^{e_1} \dots p_r^{e_n}$. Let $\zeta = \zeta_1 \dots \zeta_r$ where ζ_i is a primitive $p_i^{e_i}$-th root of 1. Adjoin the ζ_i one at a time and apply Theorem 4.13.

Let Π be a finite abelian group. Now we are going to examine the integral closure Γ, of $Z\Pi$ in $Q\Pi$. Now $Q\Pi = \Pi A_i$ where $A_i = Q(\zeta_i)$ is a cyclotonic extension of Q for all i. Thus $\Gamma = \Pi \Gamma_i$, where $\Gamma_i = Z[\zeta_i]$ is the ring of integers of $Q(\zeta_i)$. If we project $Q\Pi$ onto A_i, the image of $Z\Pi$ is $\prod_i G_0(\Gamma_i)$. Then $[M] \in G_0(Z\Pi)$.

<u>Theorem 4.16</u>. The map $G_0(\Gamma) \to G_0(Z\Pi)$ sending [M] to [M], is onto.

<u>Proof</u>. $G_0(Z\Pi) = G_0^{Z}(Z\Pi)$. Let M be a finitely generated torsion free $Z\Pi$ module. We must show [M] is in the image of $G_0(\Gamma) \to G_0(Z\Pi)$. Consider the $Q\Pi$-module $W = Q \otimes_Z M$. Choose a composition series

$$W = W_t \supset W_{t-1} \supset \dots \supset W_1 \supset W_0 = 0 \quad \text{for } W.$$

Let $M_i = M \cap W_i$. Then

$M = M_t \supset M_{t-1} \supset \dots \supset M = 0$ $M_i/M_{i-1} \subset W_i/W_{i-1}$ and M_i/M_{i-1} is torsion free. Since $[M] = \sum [M_i/M_{i-1}]$, it will suffice to consider the modules M_i/M_i, i.e., we can assume W is simple. Write $Q\pi = \prod A_i$ as above. Then W will be a module over one A_i, the remaining A_j acting trivially. Therefore $Z\pi$ acts on M through Γ_i, its image in A_i. Regard M as a Γ module with Γ_j acting trivially for $j \neq i$. Then $[M] \in G_0(\Gamma)$ maps into $[M]$ in $G_0(Z\pi)$.

<u>Theorem 4.17</u>. If π is abelian of order p^n with p a prime, then $G_0(\Gamma) \to G_0(Z\pi)$ is a isomorphism.

<u>Proof</u>. $p^n\Gamma \subset Z\pi$ by corollary 4.11. Let $S = \{1, p, p^2, p^3, \ldots\}$. Then $Z_S = Z[\frac{1}{p}]$ and $Z_S\pi = (Z\pi)_S = \Gamma_S$. Consider the commutative diagram with exact rows

$$\begin{array}{ccccccccc}
[M] & & G_0(\Gamma/p\Gamma) & \longrightarrow & G_0(\Gamma) & \longrightarrow & G_0(\Gamma_S) & \longrightarrow & 0 \\
\wr\downarrow & & \downarrow & & \downarrow & & \downarrow 1 & & \\
[M] & & G_0(Z\pi/pZ\pi) & \text{—} & G_0(Z\pi) & \text{—} & G_0((Z\pi)_S) & \text{—} & 0
\end{array}$$

where the middle map is onto by theorem 4.16. It is enough to show that $G_0(\Gamma/p\Gamma) \to G_0(\Gamma)$ is the zero map. Now $\Gamma = \Gamma_1 \times \ldots \times \Gamma_r$ where Γ_i is the ring of integers in a cyclotomic field $Q(\xi_i)$ where ξ_i is a primitive p^v-th root of 1 with $v \leq n$. We want to show that any Γ module M which is annihilated by p is zero in $G_0(\Gamma)$. Since $[M] = \sum n_i[M_i]$ where the M_i are simple $\Gamma/p\Gamma$-modules, it is enough to show that if S is a simple Γ module annihilated by p then $[S] = 0$ in $G_0(\Gamma)$.

If S simple then S is a non trivial Γ_i module for one i only. Thus it is enough to show that if $R = Z[\xi]$ where ξ is a primitive p^v-th root of 1 and S is a simple R module with pS = 0 then [S] = 0 in $G_0(R)$. Now S is certainly cyclic so we have an exact sequence $0 \to P \to R \to S \to 0$ where $p \in P$ since pS = 0. That is P occurs in the factorization of (p).

By theorem 4.13 (p) is totally ramified. $(p) = P^{\varphi(p^v)}$ and $P = (1 - \lambda)$. Therefore P is a principal ideal. But principal ideals are isomorphic to R since R is a domain. Thus [S] = [R] - [P] = 0 as desired.

Example. There exists a finite cyclic group Π such that the map $G_0(\Gamma) \to G_0(Z\Pi)$ is not an isomorphism. We shall assume without proof that there are cyclotomic fields $Q(\zeta)$ such that $Z[\zeta]$ is not a principal ideal domain. Note that $Z[\zeta]$ with ζ a primitive 23rd root of 1 is an example.

Lemma 4.18. Let k be an integer, ζ a primitive kth root of 1, and p a prime not dividing k. Then $p \mid \Phi_{pk}(\zeta)$.

Proof. Let μ be a primitive pkth root of 1 with $\zeta = \mu^p$. Then $\Phi_{pk}(x) = \Pi(x - \mu^r)$ where r runs over the classes mod pk such that (r, pk) = 1. Thus $\Phi_{pk}(\zeta) = \Pi(\mu^p - \mu^r) = (\text{unit})\Pi(1 - \mu^{r-p})$ where r runs over the classes mod pk such that (r, pk) = 1. Now μ^k is a primitive pth root of 1, $p \mid \Phi_p(1)$, and $\Phi_p(1) = \Pi(1 - \mu^{ks})$ where s runs over the classes mod p with (s, p) = 1. Thus if we can show all the factors of $\Phi_p(1)$ occur in $\Phi_{pk}(\zeta)$ we are done. That is if s is prime to p we must show there is an r prime to pk with r - p = ks. Consider all r = p + ks where s is prime to p.

Then $(r, pk) = 1$ for let q be a prime such that $q|pk$ and $q|r$. If $p = q$ then $p|r$ implies $p|ks \Rightarrow p|k$ but $p \nmid k$ by assumption. If $q \neq p$ then $q|k$ and $q|r$ so $q|p$. Thus $(r, pk) = 1$. Next we claim if we pick different classes s mod p which are prime to p, then we get different classes mod pk. If $p + ks_1 \equiv p + ks_2$ mod pk then $p|(s_1 - s_2)$. Thus every factor of $\Phi_p(1)$ occurs in $\Phi_{pk}(\zeta)$ and $p|\Phi_{pk}(\zeta)$ as desired.

Returning to the example we pick k such that $Z[\zeta]$ is not a P.I.D. where ζ is a primitive kth root of 1. Then there exists a non principal ideal $\underline{a}$ which we can choose prime to k. Say $\underline{a} = P_1 \ldots\ldots P_n$ with P_i prime. Then P_i are all prime to k and at least one is not principal. Call it P.

P must contain a prime $p \in Z$. Thus R/P has characteristic p where $p \nmid k$. Let $n = pk$ and let Π be cyclic of order n generated by x. Then $G_0(\Gamma) \to G_0(Z\Pi)$ is not a monomorphism. To see this, note $\Gamma = Z[\mu] \times Z[\zeta] \times \ldots$ where μ is a primitive n-th root of 1 and $\zeta = \mu^p$. Let $M = Z[\zeta]/P$ where P is the prime constructed above. As a $Z\Pi$ module x acts on M as ζ (i.e., $xm = \zeta m$). If $m \in M$, then $(\Phi_n(x))m = \Phi_n(\zeta)m = (\text{algebraic integer})p \cdot m = 0$.

Let $N = M$ as a module over $Z\Pi/(\Phi_n(x)) = Z[\mu]$ (considered as a factor in Γ) this makes sense since $\Phi_n(x)M = 0$. Then $[N] - [M] \in G_0(\Gamma)$ and has image 0 in $G_0(Z\Pi)$. But $[M] \neq 0$ since P is not principal and $[N] \neq [M]$ in $G_0(\Gamma)$ since they lie in different components. Therefore, $G_0(\Gamma) \to G_0(Z\Pi)$ is not an isomorphism.

We will now compute $K_0(Z\pi)$ for π cyclic of prime order p by using a theorem of Reiner which classifies all finitely generated torsion free $Z\pi$- modules. Later we will give a different proof using a theorem of Serre.

If π is cyclic of prime order p, then $Q\pi = Q \oplus Q(\zeta)$ where ζ is a primitive p-th root of 1. The maximal order Γ is $Z \oplus R$ where $R = Z[\zeta]$ is the ring of integers of $Q(\zeta)$. As above, every Z or R-module is a Γ-module through the projections $Z \oplus R \twoheadrightarrow Z$ or R and so is also a $Z\pi$-module through the inclusion $Z\pi \subset \Gamma$ (or more directly, through the projection $Z\pi \twoheadrightarrow Z$ or R).

We now single out 3 types of $Z\pi$- modules M.

<u>Type I</u> $M = Z$ (with trivial π- action)

<u>Type II</u> $M = \mathfrak{A}$, a non-zero ideal of R ($Z\pi$ acts through $Z\pi \twoheadrightarrow R$)

<u>Type III</u> Those M for which there is a non-split extension $0 \to \mathfrak{A} \to M \to Z \to 0$ where $\mathfrak{A}$ is of Type II (and Z of Type I).

We have just shown that $G_0(Z\pi) = G_0(\Gamma) = G_0(Z) \oplus G_0(R) = Z \oplus (Z \oplus C_0(R))$ where $C_0(R)$ is the class group of R. If M is a finitely generated $Z\pi$-module, we define the ideal class cl(M) of M to be the component of [M] in the summand $C_0(R)$. Clearly $0 \to M' \to M \to M' \to 0$ exact implies $cl(M) = cl(M')cl(M")$ where we consider $C_0(R)$ as a multiplicative group. Therefore $cl(M_1 \oplus \ldots \oplus M_n) = \prod_1^n cl(M_i)$. If M is of type I, cl(M) = 0 while if M has type II or III cl(M) is the ideal class of the ideal occurring in the definition of these types.

Theorem 4.19 (Reiner). Let Π be cyclic of prime order p. Then every torsion free (over Z) finitely generated $Z\Pi$-module. M is a direct sum $M = M_1 \oplus \dots \oplus M_n$ where each M_i is of type I, II, or III. Let r_1, r_2, r_3 be the number of i with M_i of type I, II, or III respectively. Then r_1, r_2, r_3 depend only on M (and not of the choice of the decomposition $M = M_1 \oplus \dots \oplus M_n$). Write $r_i(M)$ for those numbers. Then M is determined up to isomorphism by $r_1(M)$, $r_2(M)$, $r_3(M)$, and cl(M). Any set (r_1, r_2, r_3, k), $r_i \in Z$, $k \in C_0(R)$ is realized by a module M of the above type provided $r_i \geq 0$ for i = 1, 2, 3 and provided that k is trivial if $r_2 = r_3 = 0$. Finally M is projective if and only if $r_1(M) = r_2(M) = 0$.

Corollary 4.20. If Π is cyclic of prime order p, every finitely generated projective module P over $Z\Pi$ is a direct sum of modules of type III. Furthermore P is determined up to isomorphism by $rkP = pr_3(P)$ and cl(P). If $n \in Z$, $n > 0$, $p|n$ and if $k \in C_0(R)$, there is a finitely generated projective module P over $Z\Pi$ with $rkP = n$, $cl(P) = k$.

Corollary 4.21. If Π is cyclic of prime order p, $K_0(Z\Pi) = Z \oplus C_0(R)$, $G_0(Z\Pi) = Z \oplus Z \oplus C_0(R)$ and the Cartan map $\kappa: K_0(Z\Pi) \rightarrow G_0(Z\Pi)$ is given by $(n, k) \mapsto (n, n, k)$. Therefore κ is a monomorphism.

Proof. Clearly $\chi([P]) = (r_3(P), r_3(P), k)$ since this is true for P of type III. If $x \in \ker \chi$ write $x = [P] - [Q]$. Then $\chi[P] = \chi[Q]$ so $r_3(P) = r_3(Q)$, $cl(P) = cl(Q)$. Since $r_i(P) = 0 = r_i(Q)$ for $i = 1, 2$, the theorem shows $P \approx Q$ so $x = 0$. Since im χ is a group and contains all (n, n, k) for $n > 0$, $k \in C_0(R)$ we see that

$K_0(Z\pi) = \{(n, n, k) | n \in Z, k \in C_0(R)\} \approx Z \oplus C_0(R)$.

Corollary 4.22. If π is cyclic of prime order and P, Q are finitely generated projective $Z\pi$-modules with $[P] = [Q]$ in $K_0(Z\pi)$, then $P \approx Q$. More generally if X is any finitely generated $Z\pi$-module and P, Q are finitely generated torsion free $Z\pi$-modules, then $X \oplus P \approx X \oplus Q$ implies $P \approx Q$.

Proof. If T is the torsion submodule of X and $Y = X/T$, factoring out torsion shows $Y \oplus P \approx Y \oplus Q$. Since $r_i(Y \oplus P) = r_i(Y) + r_i(P)$, $cl(Y \oplus P) = cl(Y)cl(P)$ we see that $r_i(P) = r_i(Q)$, $cl(P) = cl(Q)$.

Remark. This result does not extend to all finite groups π. In fact if π is the generalized quoternion group of order 32, there is a projective $Z\pi$ module P such that $P \not\approx Z\pi$ but $Z\pi \oplus P \approx Z\pi \oplus Z\pi$.

The proof of Reiner's theorem will be based on the following well known lemma.

Lemma 4.23. Let R be a Dedekind ring, y a prime ideal of R and A a finitely generated torsion free R-module. If α is an automorphism of A/yA with det $\alpha = 1$, then α lifts to an automorphism Θ of A i.e., Θ is such that

$$\begin{array}{ccc} A & \xrightarrow[\approx]{\Theta} & A \\ \downarrow & & \downarrow \\ A/yA & \xrightarrow[\approx]{\alpha} & A/yA \end{array} \qquad \text{commutes.}$$

<u>Proof</u>. Let $A = A_1 \oplus \ldots \oplus A_n$ with A_i of rank 1. Then $A/yA = A_1/yA_1 \oplus \ldots \oplus A_n/yA_n$. Now $A_i/yA_i \approx R/y$. Let $v_i \in A_i/yA_i$ be a generator. Then $v_1, \ldots, v_n$ is a base for A/yA. With respect to this base, we can identify $\mathrm{Aut}(A/yA) = GL_n(R/y)$ and we have $\alpha \in SL_n(R/y)$. Since R/y is a field, $SL_n(R/y)$ is generated by elementary matrices in those of the form $1 + te_{ij}$ where e_{ij} has 1 in position ij and other entries 0. Let $\alpha = \varepsilon_1 \ldots \varepsilon_r$ with ε_i elementary. If each ε_i lifts to an automorphism Θ_i of A we can let $\Theta = \Theta_1 \ldots \Theta_r$. Thus it is sufficient to lift all elementary automorphisms $\varepsilon = 1 + te_{ij}$. Let $g: A_i/yA_i \to A_j/yA_j$ by $g(v_i) = tv_j$. Since A_i is projective we can lift g to $f: A_i \to A_j$ so

$$\begin{array}{ccc} A_i & \xrightarrow{f} & A_j \\ \downarrow & & \downarrow \\ A_i/yA_i & \xrightarrow[g]{} & A_j/yA_j \end{array}$$

commutes. Let φ be the endomorphism of $A = \coprod A_k$ sending A_i to A_j

by f and sending all other A_k to 0. Since $i \neq j$, $\varphi\varphi = 0$ so $(1 - \varphi)(1 + \varphi) = (1 + \varphi)(1 - \varphi) = 1$. Thus $\Theta = 1 + \varphi$ is an automorphism which clearly lifts $\mathcal{E}$.

Corollary 4.24. Let R be a Dedekind ring and y a prime ideal of R. Let $A = A_1 \oplus \ldots \oplus A_n$ be an R-module with each A_i finitely generated and torsion free of rank 1. Let $S = S_1 \oplus \ldots \oplus S_m$ be an R/y-module with $S_i \approx R/y$ for all i. If $\varphi : A \to S$ is an epimorphism, then $m \leq n$ and there is an automorphism Θ of A such that $\varphi(\Theta A_i) = S_i$ for $i \leq m$ and $\varphi(\Theta A_i) = 0$ for $i > m$.

Proof. Factor φ into $A \xrightarrow{\eta} A/yA \xrightarrow{\Psi} S$. Let s_i generate S_i. Choose $v_i \in A/yA$ with $\Psi(v_i) = s_i$. Let $\{w_j\}$ be a base for the kernel of Ψ. Then $\{v_i, w_j\}$ is a base for A/yA. We can get a new base $\{u_i\}$ for A/yA by choosing one generator u_i for each A_i/yA_i. Let α be the automorphism of A/yA sending $u_1, \ldots, u_n$ to $cv_1, v_2, \ldots, v_m, w_1, \ldots, w_h$ where we choose $c \in R/y - \{0\}$ so that $\det \alpha = 1$. By the lemma we can lift α to an automorphism Θ of A. Now $\eta(\Theta A_i) = \alpha\eta(A_i) = \alpha(A_i/yA_i) = R/yv_i$ for $i \leq m$, and R/yw_{i-m} for $i > m$. This Θ clearly has the required properties.

We now turn to the proof of Reiner's theorem. Let x generate Π. Let $N = \sum_{\sigma \in \Pi} \sigma = 1 + x + \ldots + x^{p-1} = \Phi_p(x) \in Z\Pi$. Then $Z\Pi/(N) = R$ and $Z\Pi/I = Z$ where $I = (x - 1)$ is the augmentation ideal.

Let $M^{\Pi} = \{m \in M | xm = m\}$ and $M^{N} = \{m \in M | Nm = 0\}$. Since $(x - 1)N = 0$, we have $(x - 1)M = IM \subset M^{N}$ and $NM \subset M^{\Pi}$. Therefore $A = M/M^{N}$ is a module over $Z\Pi/I = Z$ and $B = M/M^{\Pi}$ is a module over $Z\Pi/(N) = R$. These modules are torsion free (over Z) since if $h \in Z$, $h \neq 0$, $m \in M$, and $hm \in M^{\Pi}$ then $(x - 1)hm = 0$ and so $(x - 1)m = 0$ i.e., $m \in M^{\Pi}$. The same argument works for M/M^{N}. Consequently B is also torsion free over R. In fact if $r \in R$, $r \neq 0$ then r divides its norm $n \in Z$ and $n \neq 0$. Thus if $rb = 0$, $b \in B$ then $nb = 0$ so $b = 0$.

Now let $C = M/(M^{\Pi} + M^{N})$. This is a module over $Z\Pi/J$ where $J = (N) + I$. Since N maps to $p \in Z$ and I maps to $y = (\zeta - 1)$ in R we have $Z/pZ \approx Z\Pi/J \approx R/y$. We also note that $M^{N} \cap M^{\Pi} = 0$ because if m is an element of this we have $xm = m$ so $0 = Nm = pm$ but M is torsion free.

Consider the diagram

$$(\bigstar) \qquad \begin{array}{ccc} M & \xrightarrow{p} & A \\ {\scriptstyle q}\downarrow & & \downarrow{\scriptstyle r} \\ B & \xrightarrow[s]{} & C \end{array}$$

where the maps are the canonical quotient maps. This is clearly cocartesian (a pushout). This is equivalent to the exactness of $M \xrightarrow{(p,q)} A \oplus B \xrightarrow{(r,-s)} C \longrightarrow 0$. Since $\ker(p, q) = M^{N} \cap M^{\Pi}$, this sequence is also exact on the left. In other words $\bigstar$ is also cartesian. Thus we can recover M as the pullback of the diagram

(*)
$$\begin{array}{ccc} & & A \\ & & \downarrow r \\ B & \xrightarrow{s} & C \end{array}$$

Choose decompositions $A = A_1 \oplus \ldots \oplus A_a$ where $A_i \approx Z$, $C = C_1 \oplus \ldots \oplus C_c$ where $C_i \approx Z\Pi/J$ and $B = B_1 \oplus \ldots \oplus B_b$ where $B_2 \approx \ldots \approx B_b \approx R$ and $B_1 \approx \mathfrak{A}$ or a non-zero ideal of R. Applying corollary 4.24 to s (using $Z\Pi/J \approx R/y$) shows there is an automorphism θ of B such that $s(\theta B_i) = C_i$ for $i \leq c$, $s(\theta B_i) = 0$ for $i > c$. Replacing the B_i by the θB_i, we can assume $s(B_i) = C_i$ if $i \leq c$ and $s(B_i) = 0$ for $i > c$. Similarly, using $Z\Pi/J \approx Z/pZ$, we can alter the A_i so $r(A_i) = C_i$ for $i \leq c$ and $r(A_i) = 0$ for $i > c$.

Therefore the diagram * is the direct sum of diagrams of the form

(1)
$$\begin{array}{ccc} & & A_i \\ & & \downarrow \\ 0 & \longrightarrow & 0 \end{array}$$
, (2)
$$\begin{array}{ccc} & & 0 \\ & & \downarrow \\ B_i & \longrightarrow & 0 \end{array}$$
, and (3)
$$\begin{array}{ccc} & & A_i \\ & & \downarrow \\ B_i & \longrightarrow & C_I \end{array}$$

and so M is the direct sum of the pullbacks of such diagrams. The pullbacks of (1) and (2) are A_i and B_i which are of Types I and II. Consider a diagram of type (3)

$$\begin{array}{ccc} P & \xrightarrow{p} & Z \\ {\scriptstyle q}\downarrow & & \downarrow{\scriptstyle r} \\ \mathfrak{a} & \xrightarrow[s]{} & Z/pZ \end{array}$$

where P is the pullback. Since r and s are epimorphisms so are p and q. Also ker p = ker s = b say and ker q = ker r $\approx$ Z. Therefore we have exact sequences $0 \to b \to P \xrightarrow{p} Z \to 0$, $0 \to Z \to P \to \mathfrak{a} \to 0$. Therefore P is of type III unless p is a split epimorphism in which case $P \approx b \oplus Z$ is a sum of types I and II. This proves the first assertion of the theorem. We remark that, in fact, p cannot be a split epimorphism. If $h: Z \to P$ splits p then $h(Z) \subset P^{\Pi}$ so $qh(Z) \subset \mathfrak{a}^{\Pi}$. But $\mathfrak{a}^{\Pi} = 0$ since $a \in \mathfrak{a}^{\Pi}$ implies $(x - 1)a = 0$, $Na = 0$, so $pa = 0$ but $\mathfrak{a}$ is torsion free so $a = 0$. Thus $qh = 0$ so $r = rph = sqh = 0$, an obvious contradiction.

The proof of uniqueness depends on the following lemma. If R is a ring, U(R) denotes the group of units of R.

<u>Lemma 4.20</u>. $U(R) \to U(R/y)$ is onto.

<u>Proof</u>. $R/y \approx Z\Pi/J \approx Z/pZ$. If r is an integer prime to p, so that r mod p is a unit of Z/pZ, then $1 + x + \dots + x^{r-1} \in Z\Pi$ maps into r mod p. Its image in R is $\mathcal{E} = 1 + \zeta + \dots + \zeta^{r-1}$ and $\mathcal{E}$ maps into the element of R/y corresponding to r. Now $(1 - \zeta)\mathcal{E} = 1 - \zeta^{r}$. Since ζ^{r} is also a primitive p-th root of 1, $(1 - \zeta^{r})|(1 - \zeta)$ so $\mathcal{E}$ is a unit of R.

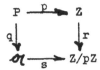

<u>Lemma 4.26.</u> Let

$$\begin{array}{ccc} P & \xrightarrow{p} & Z \\ {\scriptstyle q}\downarrow & & \downarrow{\scriptstyle r} \\ \mathfrak{A} & \xrightarrow[s]{} & Z/pZ \end{array}$$

be cartesian with r and s epimorphisms. Then the isomorphism class of P depends only on the class of $\mathfrak{A}$ (and not on r and s). Furthermore P is a projective $Z\Pi$-module.

<u>Proof.</u> We may take r(1) as the generator of Z/pZ so r is the canonical quotient map. The map s factors as

$\mathfrak{A} \to \mathfrak{A}/y\mathfrak{A} \approx R/y \xrightarrow[\Theta]{\approx} Z/pZ$. Two different maps s, s' will correspond to two different Θ, Θ'. These differ by a unit u of R/y, $\Theta' = u\Theta$. By the previous lemma, there is a unit ε of R mapping onto u. Therefore $\Theta' = \Theta\varepsilon$. This gives an isomorphism of diagrams

$$\begin{array}{ccccc} & Z & \xrightarrow{id} & Z & \\ & \downarrow{\scriptstyle r} & & \downarrow{\scriptstyle r} & \\ & Z/pZ & \xrightarrow{id} & Z/pZ & \\ {\scriptstyle s'}\nearrow & & & \nearrow{\scriptstyle s} & \\ \mathfrak{A} & \xrightarrow{\varepsilon} & \mathfrak{A} & & \end{array}$$

Since isomorphic diagrams have isomorphic pullbacks, this proves the first statement.

If we perform the construction at the beginning of the proof on $M = Z\Pi$, we see that the diagram

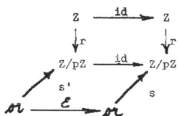

$$\begin{array}{ccc} Z\Pi & \longrightarrow & Z \\ \downarrow & & \downarrow \\ R & \longrightarrow & Z/pZ \end{array}$$

is cartesian. Given any ideal class we can choose an ideal $\mathfrak{a}$ in this class such that $R/\mathfrak{a}$ has order n prime to p. Therefore $\mathfrak{a}$ maps onto Z/pZ under the map $R \to Z/pZ$. This gives us a subdiagram

$$\begin{array}{ccc} & & Z \\ & & \downarrow \\ \mathfrak{a} & \xrightarrow[s]{} & Z/pZ \end{array}$$

with s onto. Its pullback P is a submodule of $Z\Pi$. Since $nR \subset \mathfrak{a}$, we have $nZ\Pi \subset P$. Since $p \nmid n$, Rim's lemma implies that P is projective. By the first part we get all possible isomorphism classes of P's in this way.

We can now show that M is determined by the r_1, r_2, r_3, and clM for the decomposition obtained above. We choose $B_i \approx R$ for $i \neq 1$. Therefore, the modules obtained will all be isomorphic to Z, R, or $Z\Pi$ except for the first one. This will be of type III with ideal $\mathfrak{a} = B_1$ unless $r_3 = 0$ when it will be of type II and $\approx \mathfrak{a} = B_1$. Therefore the modules occurring are determined by r_1, r_2, r_3 and the class of $\mathfrak{a}$ which is clearly cl(M).

We must now show that r_1, r_2, r_3, and clM are uniquely determined by M. Now $G_0(Z\Pi) \approx Z \oplus Z \oplus C_0(R)$ and under this isomorphism, [M] corresponds to $(r_1 + r_3, r_2 + r_3, \mathrm{cl}(M))$. Consider

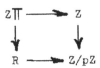

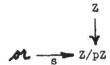

$\hat{H}^{\circ}(\Pi, M) = M^{\Pi}/N \cdot M$. If M is of type I we see that $\hat{H}^{\circ}(\Pi, M) = Z/pZ$. If M is of type II, $M^{\Pi} = 0$ so $\hat{H}^{\circ}(\Pi, M) = 0$. The same is true for M of type III by Lemma 4.7 since such M are projective. Therefore r_1 is determined by $\hat{H}^{\circ}(\Pi, M)$. Since $r_1 + r_3$, $r_2 + r_3$ are determined by M so are r_1, r_2, and r_3.

We have seen above that there is a module P of type III with any given ideal class. Therefore we can realize any $r_1, r_2, r_3 \geq 0$ and class k provided $k = 0$ if $r_2 = 0 = r_3$.

Finally we have seen that all P of type III are projective while Z is not. If $\mathfrak{A}$ is a non-zero ideal of R, let P be of type III with class $\mathfrak{A}$. We have shown above that there is an exact sequence $0 \to Z \to P \to \mathfrak{A} \to 0$. If $\mathfrak{A}$ were projective over $Z\Pi$, this sequence would split. Therefore Z would be a direct summand of P and hence would be projective. Therefore $\mathfrak{A}$ cannot be projective. This completes the proof of Reiner's theorem.

Chapter 5. Maximal Orders

Definition. Let R be a Dedekind ring, K its quotient field, and A a semisimple separable K algebra. Then an R order of A is a subring Σ of A which is a finitely generated R module such that $K\Sigma = A$.

Remark. Let M be a finitely generated torsion free R module such that $KM = A$. Let $\Lambda = \{x \in A \mid xM \subset M\}$. Then Λ is a finitely generated R module since the map $\Lambda \to \mathrm{Hom}_R(M, M)$ given by sending $\lambda \in \Lambda$ to $f \in \mathrm{Hom}_R(M, M)$ where $f(x) = \lambda x$. Embeds Λ in $\mathrm{Hom}_R(M, M)$ which is finitely generated over R. Hence, Λ is finitely generated. Clearly Λ is a ring. If $a \in A$ we can find an $r \neq 0$ in R such that $raM \subset M$ (since M is finitely generated. Therefore $K\Lambda = A$. Therefore Λ is an order called the left order of M.

Proposition 5.1. Any order can be embedded in a maximal one.

Proof. A is semisimple. Say $A = A_1 \times \ldots \times A_n$ with the A_i simple rings. Let Λ be an R order of A. Let Λ_i be the image of Λ under the projection of A to A_i. Then $\Lambda \subset \Lambda_1 \times \ldots \times \Lambda_n$ and each Λ_i is an R order of A_i. Hence $\Lambda_1 \times \ldots \times \Lambda_n$ is an R order of A. We replace Λ by $\Lambda_1 \times \ldots \times \Lambda_n$. Λ is maximal if and only if $\Lambda_i \subset A_i$ is maximal for all i. Hence we can assume A is simple. Let Z be the center of A (which is separable over K). Let R' be the integral closure of R in Z. Then $R'\Lambda$ is a finitely generated R module. Hence we can replace Λ by $R'\Lambda$ and assume K is the center of A.

Before we finish the proof we need a few lemmas.

Lemma 5.2. Let A be a central simple K algebra and L a field containing K, then $L \otimes_K A$ is a central simple L algebra.

Proof. Let e_i be a base of L over K. $L \otimes_K Z(A) \subset Z(L \otimes_K A)$ is clear (where Z(B) denotes the center of B). Let $\sum e_i \otimes a_i \in Z(L \otimes_K A)$. Let $a \in A$. Then $(1 \otimes a)(\sum e_i \otimes a_i) = (\sum e_i \otimes a_i)$. Hence $\sum e_i \otimes aa_i = \sum e_i \otimes a_i a$. But e_i is a base.

Therefore, $aa_i = a_i a$ for all $a \in A$. Hence $a_i \in Z(A)$ and $\sum e_i \otimes a_i \in L \otimes_K Z(A)$.

Next we show $L \otimes_K A$ is simple. A is simple. Hence $A = M_n(D)$, the n by n matrices over a division ring D. It is enough to show $L \otimes_K D$ is simple since a matrix ring over a simple ring is simple. Let $0 \neq B \subset L \otimes_K D$ be a proper 2 sided ideal.

Pick a non zero x in B such that the representation $x = \sum_{i \in S} e_i \otimes d_i$ has the smallest number of non zero d_i's where e_i is a base as above. Say $d_j \neq 0$. Then $(1 \otimes d_j^{-1})x = \sum_{i \in S} e_i \otimes d_j^{-1} d_i$ is in B. Hence we can assume $d_j = 1$. Pick any $d \neq 0$ in D. Then $(1 \otimes d)\, x\, (1 \otimes d^{-1}) = \sum_{i \in S} e_i \otimes dd_i d^{-1}$ is in B and has jth term $e_j \otimes 1$. Thus $(1 \otimes d)\, x\, (1 \otimes d^{-1}) - x = 0$ since it has fewer non zero terms. Therefore $d_i \in Z(D) = K$ for all i and $x \in L \otimes_K Z(D) = L$. Thus x is a unit contradicting the properness of B. Hence $L \otimes_K A$ is simple. Done with Lemma 5.2.

<u>Lemma 5.3</u>. If F is a separably closed field, D a division ring with center F, and D finite dimensional over F, then D = F.

<u>Proof</u>. Suppose not. Pick $x \in D - F$. Then F(x) is a proper purely inseparable field extension of F. Say $[F(x): F] = p^n$ where $p = \mathrm{char}(F)$. $[D: F] = [D: F(x)][F(x).F]$. Hence $p \mid [D: F]$. Also $x^{p^n} \in F$. Let $\tilde{F}$ be the algebraic closure of F. $\tilde{F} \otimes_F D \cong M_m(\tilde{F})$ and $p \mid m$ since $p \mid [D: F]$. Let $1 \otimes x$ correspond to $A \in M_m(\tilde{F})$. Now $A^{p^n} \in \tilde{F}.I$ since $(1 \otimes x)^{p^n} \in \tilde{F}$. Hence $A^{p^n} = rI = s^{p^n} I$ for some

$s \in \tilde{F}$. Hence $(A - sI)^{p^n} = A^{p^n} - s^{p^n}I = 0$ and $A - sI$ is nilpotent. Thus $tr(A - sI) = 0$ and so $tr(A) = str(I) = sm = 0$ since $p|m$. These A generate $M_m(\tilde{F})$ as an $\tilde{F}$ module. Thus everything in $M_m(\tilde{F})$ has trace 0. This is a contradiction.

Done with lemma 5.3.

<u>Definition.</u> Let A be a central simple algebra over F and $a \in A$. The <u>characteristic polynomial of a</u> is, by definition, the characteristic polynomial of $\hat{a}$: $A \to A$ given by $x \mapsto xa$.

If $E \supset F$ is field extension, it is clear that a and $1 \otimes a$ have the same characteristic polynomials.

Given A let E be the separable closure of F, then

$E \otimes_F A = M_n(E)$ by lemma 5.3. Now $M_n(E) = \coprod_1^n S$ where S is a row of length n and $\hat{a}(x) = xa$ sends these rows into themselves. Hence if f is the characteristic polynomial of a, then $f = g^n$ where g is the characteristic polynomial of $\hat{a}|S$ for a row S. Fix an isomorphism of $M_n(E)$ with $E \otimes_F A$. We will show that the coefficients of g are in F. The coefficients of f certainly are. Let s be an F automorphism of E. Since $f = g^n$ and $f = f^s = (g^s)^n$ we see that $g^n = (g^s)^n$ and so $g^s = (\text{unit})g$. All the polynomials are monic. Hence the unit = 1 and $g = g^s$. Thus the coefficients of g are fixed by s and hence are in F.

<u>Definition.</u> With the above notation g is <u>the reduced characteristic polynomial of a</u>. If $g = X^m - t(a)x^{m-1} + \dots + (-1)^m n(a)$ then

$t(a)$ is <u>the reduced trace of a</u> and $n(a)$ is <u>the reduced norm of a</u>. Clearly t is F linear and n is multiplicative.

<u>Definition</u>. Let $w_1, \ldots, w_n$ be an F base for A. Then $\Delta(w_1, \ldots, w_n) = |t(w_i w_j)|$.

If $w_i' = \sum a_{ij} w_j$ is another base, then $|t(w_i' w_j')| = |a_{ij}| |t(w_{jk})| |a_{k\ell}|$. Hence if one base has $\Delta(w_1, \ldots, w_n) \neq 0$, then every base does. But $E \otimes_F A \cong M_n(E)$. Pick the base, e_{ij}, of matrix units. Then

$$t(e_{ij} e_{k\ell}) = \begin{cases} 0 & j \neq k \text{ or } i \neq \ell \\ 1 & j = k \text{ and } i = \ell \end{cases}.$$

Thus $(t(e_{ij} e_{k\ell}))$ is a permutation matrix and has a non-zero determinant. Thus $\Delta(w_1, \ldots, w_n) \neq 0$.

<u>Lemma 5.4</u>. Let Λ be an R order of A and $a \in \Lambda$, then the reduced characteristic equation of a has all coefficients in R.

<u>Proof</u>. $R = \bigcap R_p$ where p runs over all the primes of R. Clearly $R_p \Lambda = \Lambda_p$ is an order of A over R_p. Hence we can assume R is a discrete valuation ring. Then Λ is a free R module with base $w_1, \ldots, w_n$. Now $w_i a = \sum a_{ij} w_j$ with $a_{ij} \in R$. Therefore, the characteristic equation, f, of a is in $R[x]$. But $f = g^n$ where g is the reduced characteristic equation. R is integrally closed. Therefore, g has coefficients in R. (If b_m is the highest coefficient of g with $\text{ord}_p b_m < 0$, then the coefficient of x^{nm} in f has

$\text{ord}_p < 0$. This is impossible). Done with lemma 5.4.

Now we complete the proof of proposition 5.1.

Pick a K base for A in Λ, $w_1, \ldots, w_n$. Let $w_i' = \sum a_{ij} w_j$ with $a_{ij} \in K$. Then $t(w_i w_k') = t(\sum a_{kj} w_i w_j) = \sum a_{kj} t(w_i w_j)$. But $\Delta(w_1, \ldots, w_n) \neq 0$. Hence the matrix $(t(w_i w_j))$ is nonsingular so we can pick a_{ij} such that $t(w_i w_k')$ is any given matrix over K. Pick these so that $t(w_i w_k') = \delta_{ij}$. $x \in A$, then $x = \sum t(w_i x) w_i'$. If x is in any order containing Λ (and hence containing the w_i), $t(w_i x) \in R$ by Lemma 5.4. Hence any order containing Λ is contained in $\sum R w_i'$ which is a finitely generated R module. But $\sum R w_i'$ has the ascending chain condition on R submodules. Thus we can find a maximal order containing Λ as desired. Done with proposition 5.1.

Definition. Let A be a central simple K algebra where K is the quotient field of a Dedekind ring R and let Λ be an order in A. By a (fractional) ideal of A we mean a finitely generated Λ-submodule M of A with $KM = A$. It is called integral if $I \subset \Lambda$. Define $M^{-1} = \{x \in A \mid MxM \subset M\}$. Note that $M^{-1} = \{x \in A \mid Mx \subset \Lambda\}$ since $\{y \in A \mid yM \subset M\}$ is an order containing Λ and hence equals Λ, and so $MxM \subset M \Leftarrow Mx \subset \Lambda$.

Theorem 5.5. For every left ideal M of a maximal order Λ, we have $MM^{-1} = \Lambda$.

<u>Proof</u>. If $\lambda \in \Lambda$ and $x \in M^{-1}$ then $Mx\lambda \subset \Lambda\lambda \subset \Lambda$. Hence $x\lambda \in M^{-1}$ and so M^{-1} is a right ideal of Λ. Hence MM^{-1} is a non-zero two sided ideal. To finish the proof we need several lemmas.

<u>Definition</u>. If Λ is an R order of A then a 2 sided ideal $P \subset \Lambda$ is called <u>prime</u> if for any 2 sided ideals S, T with $ST \subset P$ we have $S \subset P$ or $T \subset P$. Note that we only consider ideals with KP = A.

Let $\bar{\Lambda} = \Lambda/P$ then in $\bar{\Lambda}$, $\underline{ab} = 0$ implies $\underline{a} = 0$ or $\underline{b} = 0$ where $\underline{a}$ and $\underline{b}$ are 2 sided ideals of $\bar{\Lambda}$. $\bar{\Lambda}$ is a torsion module over R since $K \otimes_R \bar{\Lambda} = 0$. Therefore, $\bar{\Lambda}$ has D.C.C. The radical of of $\bar{\Lambda}$ is 0. In fact, the radical N is a 2 sided ideal. Since $N^m = 0$ we have $N = 0$ by the above property of $\bar{\Lambda}$. Hence, $\bar{\Lambda}$ is semisimple. If $\bar{\Lambda} = \bar{\Lambda}_1 \times \ldots \times \bar{\Lambda}_n$ with $n > 1$, then $\bar{\Lambda}_1\bar{\Lambda}_2 = 0$ which is impossible. Hence $\bar{\Lambda}$ is simple. Thus P is a maximal 2-sided ideal of Λ if and only if P is prime, i.e., P is prime if and only if L/P is simple.

<u>Lemma 5.6</u>. Any 2 sided ideal $M \subset \Lambda$ contains a product of prime ideals of Λ.

<u>Proof</u>. Suppose not. Λ is noetherian since it is finitely generated over R. Hence there is an ideal $M \subset \Lambda$ which is a maximal counterexample. Clearly M cannot be prime. Hence, there exist S and T 2 sided ideals of Λ such that $ST \subset M$. Now $(S + M)(T + M) \subset M$ so we can assume that S and T properly contain M. But then S and T contain product of primes. Thus, $ST \subset M$ does also. Done.

<u>Lemma 5.7</u>. Let S and T be (fractional) 2 sided ideals of Λ with $ST \subset \Lambda$, then $TS \subset \Lambda$.

<u>Proof</u>. $(TS)T = T(ST) \subset T\Lambda = T$. Therefore, $TS \subset$ the left order of T. This $\supset \Lambda$ and so is Λ since Λ is maximal. Hence $TS \subset \Lambda$.

Done.

<u>Lemma 5.8</u>. Let $M \subset \Lambda$ be a proper 2 sided ideal of Λ, then $M^{-1} \neq \Lambda$.

<u>Proof</u>. We first note that $KM = A$ implies $M \cap R \neq \{0\}$. Let $S = R - \{0\}$. Then $(M \cap R)_S = M_S \cap R_S = A \cap K = K$.

Now M is contained in P, a maximal 2 sided ideal by Zorn's lemma (or since Λ is noetherian). We know $P \cap R \neq \{0\}$. Pick $0 \neq u \in P \cap R$. Let $P_1 \ldots P_r \subset u\Lambda \subset P$ where the P_i are prime and r is minimal. This exists by lemma 5.6. Since P is prime, it contains some P_i and, hence, is equal to it. Changing notation we have $BPC \subset u\Lambda$ where B and C are the products of the remaining P_i. Hence $(u^{-1}B)(PC) \subset \Lambda$ where $u^{-1}B$ and PC are 2 sided ideals since u is central. Therefore, $(PC)(u^{-1}B) \subset \Lambda$ by lemma 5.7. But $P^{-1} = \{x \mid Px \subset \Lambda\}$. Hence $Cu^{-1}B \subset P^{-1}$. But $M^{-1} = \{x \mid Mx \subset \Lambda\} \supset P^{-1}$. Thus, if $M^{-1} = \Lambda$, then $P^{-1} \subset \Lambda$ so $Cu^{-1}B \subset P^{-1} \subset \Lambda$. But u is central. Thus $CB \subset u\Lambda$ and is a product of fewer than r primes. This contradicts the choice of r.

Done.

Now we finish the proof of Theorem 5.5.

Let $B = MM^{-1}$. Then B is a 2 sided Λ ideal contained in Λ. Since $\Lambda \supset BB^{-1} = MM^{-1}B^{-1}$ we have $M^{-1}B^{-1} \subset M^{-1}$. Therefore, $B^{-1} \subset$ right order of $M^{-1} = \Lambda$ since Λ is a maximal order. Therefore, $B = \Lambda$ by the previous lemma.

Done.

Corollary 5.10. Let Λ be a maximal order and M a 2 sided Λ-ideal, then M is projective as a left or right Λ module. (This will be considerably improved in Theorem 5.12.)

Proof. We have $M^{-1}M = \Lambda$ since M is a right ideal by Theorem 5.5 (applied to the opposite ring of Λ i.e., Λ with $x \bullet y = yx$). But the definition of M^{-1} is independent of left or right. Hence $M^{-1}M = \Lambda$ for 2 sided ideals. Pick $m_i \in M$ and $n_i \in M^{-1}$ such that $\sum_{i=1}^{r} n_i m_i = 1$. We map $A \xrightarrow{i} \underbrace{\Lambda \oplus \cdots \oplus \Lambda}_{r \text{ times}}$ by $i(x) = (xn_1, \ldots, xn_r)$ and $\Lambda \oplus \cdots \oplus \Lambda \xrightarrow{j} M$ by $j(k_1, \ldots, k_r) = k_1 m_1 + \cdots + k_r m_r$. Then $ji(x) = \sum_{i=1}^{r} sn_i m_i = x(\sum_{i=1}^{r} n_i m_i) = x$. Hence M is projective as a left module since it is a direct summand of $\Lambda \oplus \cdots \oplus \Lambda$. Similarly M is projective as a right module. Done.

Theorem 5.11. Let Λ be a maximal order in a central simple algebra A over K. Then the 2 sided (fractional) ideals form a free abelian group with generators the prime ideals of Λ.

Proof. 0) The 2 sided ideals clearly form a monoid under multiplication. The unit is Λ. This monoid is a group since $MM^{-1} = M^{-1}M = \Lambda$ for all M in it.

1) Let $M \subset \Lambda$ be a 2 sided ideal. We first show that M is the product of primes of Λ. Suppose not. Let M be a maximal

counter example. Now $M \subset P$ for some prime P so $P^{-1}M \subset P^{-1}P = \Lambda$. Also $M \subsetneq P^{-1}M$ since $M = P^{-1}M$ would imply $\Lambda = MM^{-1} = P^{-1}MM^{-1} = P^{-1}$ and so $P = P\Lambda = PP^{-1} = \Lambda$. Thus $P^{-1}M$ is a product of primes by the maximality of M. But $M = P(P^{-1}M)$.

2) If M is a fractional ideal, then M is the product of primes and their inverses. In fact, there exists $u \in R$, $u \neq 0$ with $uM \subset \Lambda$ since M is finitely generated. Let $B = uM$. Then B is a proper 2 sided ideal. Hence, $B = P_1 \ldots\ldots P_r$, $u\Lambda = P_1' \ldots\ldots P_s'$ and $M = P_s'^{-1} \ldots\ldots P_1'^{-1}P_1 \ldots\ldots P_r$.

3) The group is abelian. Let P and P' be primes. Then $P' \supset P'P = P(P^{-1}P'P)$ so either $P \subset P'$ or $P^{-1}P'P \subset P'$. (Note that $P'P \subset P$ so $P^{-1}P'P \subset P^{-1}P = \Lambda$). If $P \subset P'$, then $P = P'$ since primes are maximal 2 sided ideals and so P and P' commute. If $P \neq P'$, then $P^{-1}P'P \subset P'$ so $P'P \subset PP'$. By symmetry $PP' \subset P'P$ i.e., $P'P = PP'$. Thus all generators commute.

4) The group is free abelian. Let

$P_1^{u_1} \ldots P_r^{u_r}P_1'^{-v_1} \ldots P_s'^{-v_s} = \Lambda$ be a relation with all $u_i, v_j > 0$. Then $P_1^{u_1} \ldots\ldots P_r^{u_r} = P_1'^{v_1} \ldots P_s'^{v_s} \subset P_1'$. Therefore, since P is prime we have $P_i \subset P_1'$ and so $P_j = P_1'$ for some i. Multiplying by $P_1'^{-1}$ gives $P_1^{u_1} \ldots P_i^{u_i-1} \ldots = P_1'^{v_1-1} P_2^{v_2} \ldots$. By induction on $\sum u_i$ we see that all relations are trivial. Done.

<u>Theorem 5.12</u>. Let Λ be a maximal order in a central simple K algebra and M be a finitely generated torsion free left Λ module. Then M is projective.

<u>Proof</u>. Consider $K \otimes_R M$ as an A module. Since A is simple, every A module is projective. Therefore $(K \otimes_R M) \oplus (K \otimes_R N) = A^m$ for some integer m and some finitely generated torsion free module N. If $M \oplus N$ is projective, then M is. Hence, using $M \oplus N$ for M, we can assume $K \otimes_R M$ is free and even that $K \otimes_R M = A^m$, i.e., $M \subset A^m$ and $KM = A^n$. Since M is finitely generated, there is an $r \neq 0$ in R with $rM \subset \Lambda^m \subset A^m$. Since M torsion free, rM is isomorphic to M. Hence we can assume that $M \subset \Lambda^m$ and $KM = K\Lambda^m = A^m$. Therefore, Λ^m/M is a torsion module over R. Thus, considering the resolution $0 \to M \to \Lambda^m \to \Lambda^m/M \to 0$, it is enough to show the following lemma.

<u>Lemma 5.13</u>. If B is an Λ module which is finitely generated and torsion over R, then the projective dim over Λ of B is ≤ 1.

<u>Proof</u>. B has a composition series, $B = B_0 \supset B_1 \supset \dots \supset B_t = 0$. Consider $0 \to B_1 \to B \to B/B_1 \to 0$. If $pd_\Lambda B_1 \leq 1$ and $pd_\Lambda B/B_1 \leq 1$, then $pd_\Lambda B \leq 1$. Thus, by induction on t we can assume B is a simple Λ module. Since B is torsion, and finitely generated, there is some $r \neq 0$ in R with $rB = 0$. Now $r\Lambda$ is a 2 sided ideal. Hence, $r\Lambda = P_1^{v_1} \dots P_s^{v_s}$ with P_i primes. By the simplicity of B, if I is a 2 sided ideal either $IB = 0$ or $IB = B$. Since $(r\Lambda)B = 0$, there must exist a prime $P = P_i$ with $PB = 0$.

Therefore, $B = B/PB$ and B is an Λ/P module. But Λ/P is simple so B is a direct summand of a free Λ/P-module. Hence, it is enough to show that $pd_\Lambda(\Lambda/P) \leq 1$. But $0 \to P \to \Lambda \to \Lambda/P \to 0$ is exact and P is projective by corollary 5.10. Done.

Corollary 5.14. Let Λ be a maximal order in a central simple K algebra A. Then every finitely generated Λ-module M has $pd_\Lambda M \leq 1$. Thus Λ is left (and right) regular so $K_0(\Lambda) = G_0(\Lambda)$.

Proof. Let M be a finitely generated module. Let $0 \to P \to F \to M \to 0$ with F free. Then P is projective by Theorem 5.12. The last part follows from Theorem 1.1.

Problem: Is corollary 5.14 true for non finitely generated modules?

Theorem 5.15. Let A be a central simple K algebra with Λ and Γ maximal orders. Then the categories of left Γ and left Λ modules are equivalent and this equivalence restricts to one between the subcategories of finitely generated modules. A Γ module is torsion free (resp. torsion) if and only if the corresponding Λ-module is.

Proof. This is again a special case of the Morita theorems. Let $B = \Lambda\Gamma$. This is a finitely generated R submodule of A which is a left Λ and right Γ ideal. Let $C = B^{-1}$. Then C is a left Γ and right Λ ideal. If M is a left Λ module send it to $C \otimes_\Lambda M$. If N is a left Γ module send it to $B \otimes_\Gamma N$. Then the composite sends M to $B \otimes_{\Gamma} C \otimes_{\Lambda} M$ which we claim is isomorphic to to $\Lambda \otimes_{\Lambda} M = M$. We only need to show that $B \otimes_{\Gamma} C \to BC = \Lambda$ is an isomorphism. The mapping clearly is onto. Applying $K \otimes_R$— we see that the kernel of $B \otimes_{\Gamma} C \to \Lambda$ is torsion. Now B is a projective right Γ module since it is finitely generated and torsion free. Thus there exists a right Γ module Y such that $B \oplus Y = \Gamma^s$. Then $(B \otimes_{\Gamma} C) \oplus (Y \otimes_{\Gamma} C) = (B \oplus Y) \otimes_{\Gamma} C = C^s$ is torsion free. Therefore, $B \otimes_{\Gamma} C$ is torsion free. Hence the kernel of $B \otimes_{\Gamma} C \to BC$ is 0. Similarly $C \otimes_{\Lambda} B$ is isomorphic to Γ. Hence, the composition is also the identity. The functors $C \otimes_{\Lambda}$ —, $B \otimes_{\Gamma}$— give the required equivalence. If M is finitely generated (resp torsion) so is $C \otimes_{\Lambda} M$. If M is torsion free, then for all $r \in R$, $r \neq 0$ we have $0 \to M \xrightarrow{r} M$. Since C is projective over Λ, we have $0 \to C \otimes_{\Gamma} M \xrightarrow{r} C \otimes_{\Gamma} M$ so $C \otimes_{\Gamma} M$ is torsion free.

Theorem 5.16. Let $p \subset R$ be a prime ideal $\neq 0$. Then there exists a unique prime P of Λ such that $P \supset p\Lambda$. As always Λ is a maximal order in a central simple K-algebra.

Proof. Let $\hat{\Lambda} = \varprojlim \Lambda/p^n\Lambda$, the completion of Λ at p and let $\hat{R} = \varprojlim R/pR$. Then $\hat{R} \otimes_R \Lambda = \hat{\Lambda}$ since Λ is a finitely generated R-module. If $p\Lambda = P_1^{v_1} \dots P_s^{v_s}$, then $(p\Lambda)^n = P_1^{nv_1} \dots P_s^{nv_s}$

since the primes commute. Thus, $\Lambda/p^n\Lambda = \prod_{i=1}^{s} \Lambda/P_i^{nv_i}$ by the Chinese Remainder Theorem which holds since 2 sided ideals commute. Let $e_i^{(n)}$ be the identity of $\Lambda/P_i^{nv_i}$. Then $e_i = (\ldots \to e_i^{(n+1)} \to e_i^{(n)} \to \ldots) \in \hat{\Lambda}$ and is an idempotent. Thus $\hat{\Lambda} = \prod_{i=1}^{s} \Lambda e_i$. Let $\hat{K}$ be the quotient field of $\hat{R}$. Then $\hat{K} \otimes_{\hat{R}} \hat{\Lambda} = \prod_{i=1}^{s} \hat{K} \otimes_{\hat{R}} (\hat{\Lambda} e_i) = \hat{K} \otimes_K A$ is central simple over $\hat{K}$ by lemma 5.2. Thus s = 1 and there is only one prime over each prime of R.

Done.

<u>Proposition 5.17</u>. Let R be a (not necessarily commutative ring) and let A be a nilpotent 2 sided ideal, then if $e \in R/A$ is an idempotent, there exists an idempotent $e' \in R$ with image e in R/A.

<u>Proof</u>. By induction on n with $A^n = 0$, we can assume $A^2 = 0$ i.e., lift e to R/A^2 and from there to R. Let $x \in R$ be any preimage of e and let $a = x^2 - x$. Then $a \in A$ since e is idempotent. Let $y = (x - a)^2 = x^2 - 2ax = x + a - 2ax$. Then y is the desired element since $y^2 = x^2 + 2x(a - 2ax) = x + a + 2ax - 4a(x + a) = x + a - 2ax = y$.

<u>Corollary 5.18</u>. Let R be a complete local ring with maximal ideal p and Λ an R algebra which is finitely generated as an R module. If e is an idempotent of $\Lambda/p\Lambda$, there is an idempotent e' of Λ mapping onto it.

Proof. Λ is isomorphic to $\varprojlim \Lambda/p^n\Lambda$ since R is complete. But $p\Lambda/p^n\Lambda$ is a nilpotent ideal of $\Lambda/p^n\Lambda$. Hence by Prop. 5.17 we can lift e to an element e_n in each $\Lambda/p^n\Lambda$ such that the liftings are compatible. Thus, we lift e to $(e_n) \in \varprojlim \Lambda/p^n\Lambda$.

Corollary 5.19. Let R be a complete local ring with maximal ideal p and Λ an R algebra which is a finitely generated R module and which has no idempotents except 0 and 1. Then Λ is local.

Proof. Let $x \in \Lambda$ and let $\bar{x}$ be its image in $\Lambda/p\Lambda$. Then x is a unit if and only if $\bar{x}$ is. For if multiplication by x is an epimorphism mod $p\Lambda$ then it is an epimorphism by Nakayama's lemma. Thus x has right and left inverses if $\bar{x}$ does. Hence x is a unit if $\bar{x}$ is. The converse is obvious. Next let J be the Jacobson radical of $\Lambda/p\Lambda$. Since $\Lambda/p\Lambda$ is finite dimensional over R/p, J is nilpotent. Hence $\bar{x}$ is a unit if and only if its image $\bar{\bar{x}} \in \Lambda/J$ is. By assumption Λ has no idempotents except 0 and 1. Hence, Λ/J has no idempotents except 0 and 1 by corollary 5.18. Thus Λ/J is a division ring, and by the above results, every element of Λ - J is a unit of Λ. Thus J is the unique maximal ideal of Λ.

Theorem 5.20. Let R be a complete local ring and Λ an R algebra. Then the Krull-Schmidt Theorem holds for Λ modules finitely generated over R.

Proof. By SK, Theorem 2.11 we need to prove two things.

1) Every finitely generated Λ module is a sum of indecomposable Λ modules, and

2) If M is an indecomposable Λ module, then $E = \text{Hom}_{\Lambda}(M, M)$ is local.

1) follows since finitely generated Λ modules are noetherian.

2) $E \subset \text{Hom}_R(M, M)$ which is finitely generated over R. Hence E is finitely generated over R. E has no idempotents except 0 and 1 since an idempotent endomorphism e gives a decomposition $M = \ker e \oplus \operatorname{im} e$. Hence, E is local by corollary 5.19.

<u>Corollary 5.21</u>. With the above hypothesis $K_0(\Lambda)$ is a free abelian group generated (freely) by the classes of the indecomposable projective modules.

<u>Definition</u>. Let A and B be R modules and $f: A \to B$ a homomorphism. Then f is an <u>essential epimorphism</u> if f is an epimorphism and if X is any R module and $g: X \to A$ is any homomorphism such that fg is an epimorphism. Then g is an epimorphism.

<u>Definition</u>. A <u>projective cover</u> of an R module M is a projective R module P with an essential epimorphism $f: P \to M$.

<u>Remark</u>. If $f: P \to M$ and $f': P' \to M$ are two projective covers, then P and P' are isomorphic. Since P is projective we can find $\theta: P \to P'$ with $f'\theta = f$. Since f' is essential, θ is an epimorphism. Since P' is projective there is a $\varphi: P' \to P$ so $\theta\varphi = \text{id}$. Since f is essential, φ is an epimorphism so θ and φ are isomorphisms.

Lemma 5.22. Let R be a (not necessarily commutative) ring, A a nilpotent 2 sided ideal, and M an R module. Then $p: M \to M/AM$ is an essential epimorphism.

Proof. Let $g: X \to M$ be a map with pg an epimorphism. Let $g(X) = N$. Since pg is an epimorphism $M = N + AM$. Hence $M/N = A(M/N) = A^2(M/N) = \ldots = A^n(M/N) = 0$. Thus, $M = N$ and g is an epimorphism.

Lemma 5.23. Let A be a nilpotent 2 sided ideal of a ring R and Q a finitely generated projective R/A module. Then Q has a projective cover over R. In fact, Q is isomorphic to P/AP where P is a finitely generated projective R module.

Proof. There exists a Q' such that $Q \oplus Q' = (R/A)^n = F$ and an idempotent matrix $e: F \to F$ with $Q = e(F)$. The sequence

$$0 \to M_n(A) \to M_n(R) \to M_n(R/A) \to 0$$

is exact when $M_n(B) = \{n \times n \text{ matrices with entries in } B\}$. Clearly $M_n(A)$ is a nilpotent ideal of $M_n(R)$. Hence, e lifts to e', an idempotent of $M_n(R)$. Let $P = \mathrm{im}(e')$. Now $Q = F/(1 - e)F$ and $P = R^n/(1 - e')R^n$. Tensoring $R^n \xrightarrow{(1-e')} R^n \longrightarrow P \longrightarrow 0$ with R/A gives $(R/A)^n \xrightarrow{(1-e)} (R/A)^n \longrightarrow P/PA \longrightarrow 0$ so $P/PA \approx Q$.

Proposition 5.24. Let A be a nilpotent ideal of a ring R. Then there is a 1 - 1 correspondence between the isomorphism classes of finitely generated projective R modules and finitely generated projective R/A modules giving by passing from P to P/AP and from Q to a projective cover of Q as an R module. This correspondence induces an isomorphism of $K_0(R)$ with $K_0(R/A)$.

Proof. Immediate.

Remark. The same theorem is true for the case in which A is not nilpotent but R is complete with respect to A, i.e., $R = \varprojlim R/A^n$.

We lift idempotents by the argument of corollary 5.18.

Corollary 5.24A. Let R be a complete local ring and Λ an R-algebra finitely generated as an R-module. Then there are only a finite number of indecomposable projective Λ modules. Hence $K_0(\Lambda)$ is finitely generated.

Proof. If J is the Jacobson radical of Λ then Λ is complete with respect to J.

Proposition 5.25. If R is an artinian ring, then every finitely generated R module has a projective cover.

Proof. Let $\underline{r}$ = radical(R). $\underline{r}^n = 0$ for some n since R is artinian. Let M be a finitely generated R module. Then $M \twoheadrightarrow M/\underline{r}M$ is an essential epimorphism and $R/\underline{r}$ is semisimple. Therefore, $M/\underline{r}M$ is projective as an $R/\underline{r}$ module. Hence, $M/\underline{r}M$ has a projective cover $P \twoheadrightarrow M/\underline{r}M$ as an R module P is projective and $M \twoheadrightarrow M/\underline{r}M$ is an epimorphism. Hence, there exists an $f: P \twoheadrightarrow M$ making

$$\begin{array}{ccc} & & P \\ & \overset{f}{\swarrow} & \downarrow \\ M & \xrightarrow[\eta]{} & M/\underline{r}M \end{array}$$

commute.

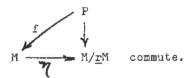

Then f: P $\twoheadrightarrow$ M is a projective cover. To see that f is essential note that if fg is an epimorphism so is ηfg. Therefore so is g. Since η itself is essential, f must be an epimorphism.

Corollary 5.26. Let R be an artinian ring. Then $K_0(R)$ and $G_0(R)$ are isomorphic (but not necessarily by the Cartan map).

Proof. $K_0(R)$ is free on indecomposable projectives. $G_0(R)$ is free on simple modules and these bases are in 1 - 1 correspondence by factoring out by the radical and taking projective cover.

Theorem 5.27. If A is a separable semisimple algebra over K, the quotient field of a discrete valuation ring R, Λ is a maximal order over R in A, and M and N finitely generated torsion free Λ modules with KM isomorphic to KN, then M is isomorphic to N. In particular every ideal of Λ is principal.

Proof. By Theorem 1.10, it will suffice to show that the Cartan map $K_0(\Lambda) \twoheadrightarrow G_0(\Lambda)$ is a monomorphism. Let $A = A_1 \times \dots \times A_n$ with the A_i simple. Then $\Lambda = \Lambda_1 \times \dots \times \Lambda_n$ with Λ_i a maximal order in A_i. Then $\Lambda/p\Lambda = \prod \Lambda_i/pA_i$ where p is the maximal ideal of R. G_0 and K_0 preserve finite sums. Hence it is enough to consider the case A simple. Let K' be its center and R' the integral closure of R in K'. Then Λ is a maximal order over R' as well as over R. Now pR' is an ideal of R'. Say

Say $pR' = p_1^{v_1} \dots p_r^{v_r}$ with the p_i primes of R'. Then $\Lambda/p\Lambda = \prod \Lambda/p_i^{v_i}\Lambda$ and it remains to check that the Cartan map is a monomorphism for the factors $\Lambda/p_i^{v_i}\Lambda$. By Theorem 5.16, there is a unique P in Λ containing $p_i\Lambda$. Thus $\Lambda/p_i^{v_i}\Lambda$ has a unique maximal 2 sided ideal so $(\Lambda/p_i^{v_i}\Lambda)$/radical is simple. Therefore $\Lambda/p_i^{v_i}\Lambda$ has only one simple module S and so only one indecomposable projective P. Thus $G_0(\Lambda/p_i^{v_i}\Lambda) = Z$ generated by by [S] and $K_0(\Lambda/p_i^{v_i}\Lambda) = Z$ generated by [P]. The Cartan map is multiplication by n, the number of times S occurs in a composition series for P. This map is clearly a monomorphism.

Corollary. Theorem 5.27 remains true if R is any semilocal Dedekind ring.

Proof. If $p_1, \dots, p_n$ are the primes of R. Then R_{p_i} is a discrete valuation ring and Λ_{p_i} is a maximal order over R_{p_i} by Theorem 5.28 below. Since $KM_{p_i} = KM$ isomorphic to $KN_{p_i} = KN$, Theorem 5.27 implies M_{p_i} is isomorphic to N_{p_i} say by f_{p_i}, i.e., M and N have the same genus. By Roiters lemma we can find $0 \to M \to N \to X \to 0$ where the annihilator of X is prime to $p_1, \dots, p_r$. Therefore $X = 0$.

Theorem 5.28. Let R be a commutative noetherian ring, and Λ a maximal order over R in a separable K algebra A where K is the quotient field of R. Let $S \subset R$ be a multiplicatively closed set. Then Λ_S is a maximal order over R_S in A.

Proof. Suppose not. Let $\Lambda_S \subset \Gamma$ where Γ is an R_S order in A. Then there exists $r \in R$ such that $F = r\Gamma \subset \Lambda_S$. Then $\underline{f} = F \cap \Lambda$ is a 2 sided ideal of Λ. Let $O_\chi(f) = \{x | x\underline{f} \subset \underline{f}$ and $x \in A\}$. Then $O_\chi(\underline{f}) \supset \Lambda$ and is an order over R. Therefore, $O_\chi(\underline{f}) = \Lambda$ since Λ is maximal. Pick $\gamma \in \Gamma$. Then $\gamma F \subset F$ and $\gamma \underline{f} \subset \gamma F \subset F = r\Gamma \subset \Lambda_S$. $\underline{f}$ is a finitely generated R module since it is contained in Λ. Therefore, there exists $s \in S$ such that $s\gamma \underline{f} \subset \Lambda$ and $s\gamma \underline{f} \subset sF \subset F$. Therefore, $s\gamma \underline{f} \subset \Lambda \cap F = \underline{f}$. Hence $s\gamma \in O_\chi(\underline{f}) = \Lambda$ and so $\gamma = \frac{s\gamma}{s} \in \Lambda_S$. Hence $\Gamma = \Lambda_S$, and Λ_S is a maximal R_S order as desired.

Corollary 5.29. Let A be a separable semisimple K algebra where K is the quotient field of a noetherian ring R. Let Λ be an order in A over R. Then Λ is a maximal order if and only if Λ_p is a maximal order over R_p for every maximal ideal p of R.

Proof. Embed Λ in Γ a maximal order. Then $\Lambda_p \subset \Gamma_p$. If all Λ_p are maximal orders, then $\Lambda_p = \Gamma_p$ for all p and so $\Lambda = \Gamma$. The converse is a special case of Theorem 5.28.

Chapter 6. Orders

In this chapter we present results of Jacobinski and Roiter concerning arbitrary orders. Let A be a separable semisimple K algebra where K is the quotient field of a Dedekind ring R. Let Λ be any R order in A and let $\Gamma \supset \Lambda$ be a maximal order. Then there exists an $r \neq 0$ in R with $r\Gamma \subset \Lambda$. If p is a prime ideal of R with $r \notin p$, then $r\Gamma_p \subset \Lambda_p \subset \Gamma_p$ and r is a unit at p. Hence, $\Lambda_p \supset \Gamma_p$. There are only a finite number of primes of R containing r. Thus, except for those primes, Λ_p is a maximal order.

Lemma 6.1. Let M be a finitely generated torsion free Λ module, $B \to C \to 0$ an exact sequence of Λ modules and $f: M \to C$ a Λ homomorphism. Then rf lifts to a map, g, from M to B making the diagram

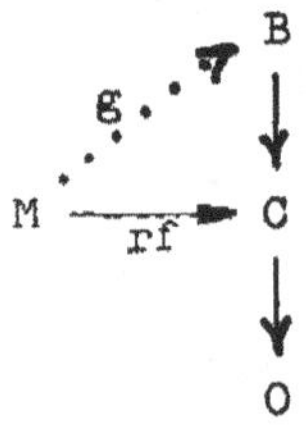

commute.

Corollary 6.2. $r\,Ext^1_\Lambda(M, X) = 0$ for all Λ modules X.

Proof of corollary from the lemma. Let $0 \to X \to I \to Y \to 0$ be exact with I an injective Λ module. Then $0 \to Hom_\Lambda(M,X) \to Hom_\Lambda(M,I) \to Hom_\Lambda(M,Y) \to Ext^1_\Lambda(M,X) \to 0$ is exact. Pick $u \in Ext^1_\Lambda(M, X)$. Lift u to $f \in Hom_\Lambda(M, Y)$. Then rf is a lifting of ru and rf comes from some $g \in \mathrm{Hom}_\Lambda(M, I)$ by the lemma. Hence, rf goes to zero in $\mathrm{Ext}^1_\Lambda(M, X)$. Hence $ru = 0$.

<u>Remark</u>. Actually we have shown more. If $\underline{a} = \{r \in R \mid r\Gamma \subset \Lambda\}$, then $\underline{a}\,\mathrm{Ext}^n_\Lambda(M, X) = 0$ for $n > 0$. This follows by exactly the same method.

<u>Proof of the lemma.</u> Let F be a finitely generated free module mapping onto M. Let $\hat{r}: M \to M$ be given by multiplication by r. Since F is free there is a map, h, making

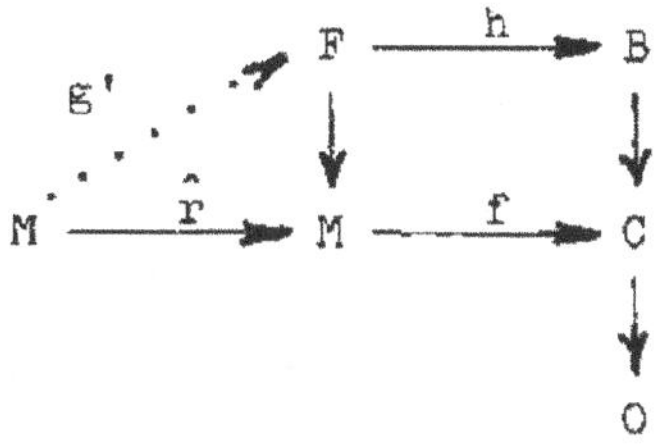

commute.

If we can find a $g': M \to F$ making the diagram commute, then hg' will lift rf. Hence we can assume $C = M$ and B is a finitely generated free module mapping onto C. If we were over Γ we would be done by Theorem 5.12. To take advantage of this, we tensor with Γ and obtain the diagram

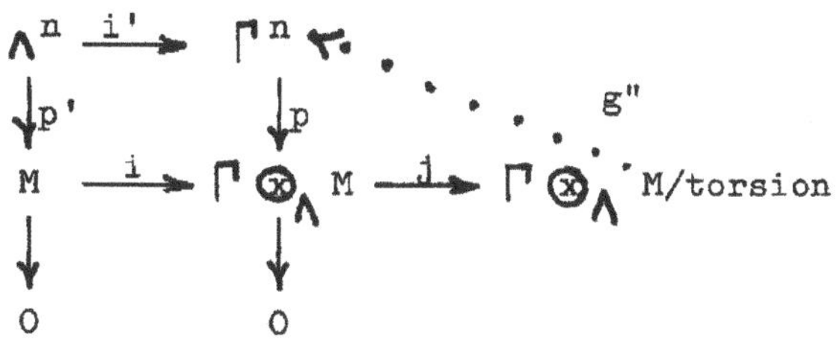

$\Gamma \otimes_\Lambda M/\text{torsion}$ is Γ projective and since pj is onto, there is a g" making the diagram commute. The image of rg" will be contained in $r\Gamma^n \subset \Lambda^n$. Hence rg"ji will lift $\hat{r}$ to a map $M \to \Lambda^n$. To check that $p'(rg''ji) = \hat{r}$ we can apply ji which is injective since M is torsion free.

We recall the following

<u>Definition</u>. Let M and N be finitely generated torsion free Λ modules where Λ is an order over R. Then we say M and N have the same genus, $(M \sim N)$, if M_p is isomorphic to N_p over Λ_p for all primes p of R. This terminology will only be applied to finitely generated torsion free Λ-modules.

<u>Lemma 6.3</u>. Let R be a discrete valuation ring with maximal ideal p, Λ an R order, M and N finitely generated Λ modules, $\hat{R}$ the completion of R at p, $\hat{\Lambda} = \Lambda \otimes_R \hat{R}$, $\hat{M} = M \otimes_R \hat{R}$, and $\hat{N} = N \otimes_R \hat{R}$. Then M is isomorphic to N if and only if $\hat{M}$ is isomorphic to $\hat{N}$.

<u>Proof</u>. Let $g\colon \hat{M} \to \hat{N}$ be an isomorphism. $\hat{R}$ is a flat R module and M and N are finitely presented. Hence, $\hat{R} \otimes_R \mathrm{Hom}_\Lambda(M, N)$ is isomorphic to $\mathrm{Hom}_{\hat\Lambda}(\hat{M}, \hat{N})$. Thus, there exists an $f \in \mathrm{Hom}_\Lambda(M, N)$ such that $1 \otimes f \equiv g \bmod p$. Hence $g - (1 \otimes f)\colon \hat{M} \to p\hat{N}$. Consider

$$\begin{array}{ccc} M & \xrightarrow{f} & N \\ \downarrow & & \downarrow \\ M/pM & \xrightarrow{f^*} & N/pN \end{array} \qquad \text{and} \qquad \begin{array}{ccc} \hat{M} & \xrightarrow[g]{1 \otimes f} & \hat{N} \\ \downarrow & & \downarrow \\ \hat{M}/p\hat{M} & \xrightarrow[g^*]{(1 \otimes f)^*} & \hat{N}/p\hat{N} \end{array} .$$

Since g is an isomorphism and $1 \otimes f \equiv g \bmod p$, we see that $(1 \otimes f)^*$ is onto. But $(1 \otimes f)^* \cong f^*$ under the natural identification of M/pM with $\hat{M}/p\hat{M}$ and N/pN with $\hat{N}/p\hat{N}$. Thus f^* is onto. Hence f is onto by Nakayama's lemma. Since rank N = rank M the kernel is torsion. But M is torsion free. Hence f is an isomorphism.

The converse is obvious.

We return now to an order Λ over a Dedekind ring R, in a semisimple K-algebra.

<u>Lemma 6.4</u>. If M, M', N, and N' are Λ modules, finitely generated and torsion free with $M \oplus N \sim M' \oplus N'$ and $M \sim M'$, then $N \sim N'$.

<u>Proof</u>. By lemma 6.3 it is enough to check at the completion. But the Krull-Schmidt theorem holds for orders over complete local rings.

<u>Remark</u>. In checking whether $M \sim N$ we can restrict our attention to a finite set of primes of R by the following method. Let Γ be a maximal order containing Λ. Then there is an $r \neq 0$ in R with $r\Gamma \subset \Lambda$. Let $\underline{P}$ be a finite non empty set of primes of R including all the primes containing r.

<u>Lemma 6.5</u>. Two finitely generated torsion free Λ modules M and N will have the same genus if and only if N_p is isomorphic to M_p for all $p \in \underline{P}$.

<u>Proof</u>. If K is the quotient field of R, then $K \otimes_R M$ is isomorphic to $K \otimes_R N$ since $\underline{P}$ is non empty. Now M_p and N_p $p \notin P$ are finitely generated torsion free modules over a maximal R_p order $\Lambda_p = \Gamma_p$

which become isomorphic over K. Therefore, by theorem 5.27, M_p is isomorphic to N_p. Hence M and N are isomorphic at every prime of R if they are so at all $p \in \underline{P}$.

Recall now the following definition.

Definition. Let R be a commutative domain. Then the Jordan-Zassenhaus Theorem holds for R if for every R algebra Λ which is finitely generated and torsion free as an R module with $K \otimes_R \Lambda$ semisimple and separable over K (K is the quotient field of R) and for every integer $n \geq 0$ there are only a finite number of isomorphism classes of Λ modules which are finitely generated and torsion free of rank $\leq n$.

Lemma 6.6 (Roiter). Let R be a Dedekind ring such that the Jordan Zassenhaus Theorem holds for R. Let Λ be an order in a semisimple separable K algebra A. Let $0 \neq \underline{a}$ be an ideal of R. Let M and N be two Λ modules of the same genus. Then there exists an exact sequence $0 \to M \to N \to U \to 0$ where U is torsion free with annihilator prime to $\underline{a}$ and with $U = \coprod U_i$ where the U_i are simple Λ modules and the annihilators of the U_i are pairwise relatively prime.

Proof. Let n = rank M. Look at all Λ modules M_i of rank n which are finitely generated and torsion free and all exact sequences $0 \to M_i \xrightarrow{f} M_j \to S_f \to 0$ where the S_f is simple, and, hence, its annihilator is a prime ideal p. For each pair of integers i, j one of the following happens

1) Infinitely many primes of R are obtainable as annihilators of S_f for differing f.

2) Only finitely many primes can occur as annihilators of S_f.

Since there are only finitely many pairs i, j, only finitely many occur in case 2). Let $\underline{b}$ be the product of all the primes that can occur in 2) for any i, j. Apply Roiter's theorem to the ideal $\underline{a}\,\underline{b}$ to obtain an exact sequence $0 \to M \to N \xrightarrow{j} X \to 0$ where X has annihilator prime to $\underline{a}\,\underline{b}$. Let $X = X_0 \supset \ldots \supset X_m = 0$ be a composition series for X and $N_i = j^{-1}(X_i)$. Then $N = N_0 \supset N_1 \supset \ldots \supset N_{m-1} \supset N_m = M$ has all quotients N_i/N_{i+1} simple. Consider $N_0 \supset N_1$. For some i, j N_0 isomorphic to M_j and N_1 isomorphic to M_i and $0 \to N_1 \to N_0 \to S \to 0$ is exact with the annilator of S prime to $\underline{a}\,\underline{b}$. Hence, by the choice of $\underline{b}$, case 1) applies to (i, j). Hence we can modify the inclusion $N_1 \hookrightarrow N_0$ so that N_0/N_1 is simple with annihilator any of an infinite number of primes. Doing this at each stage we get a new embedding of M into N with $0 \to M \to N \to U \to 0$ exact and $U = U_0 \supset U_1 \supset \ldots \supset U_n = 0$ a composition series with the annihilators of U_{i-1}/U_i pairwise relatively prime. Then U is the direct sum of its primary components which are simple Λ modules isomorphic to the U_{i-1}/U_i. This is what we wanted.

Lemma 6.7. (Roiter). Let R be any Dedekind ring. Let Λ be an R order in a semisimple separable K algebra A, Γ a maximal order containing Λ, $r \neq 0$ in R with $r\Gamma \subset \Lambda$. Let U be a Λ module which is the direct sum of $U_1, \ldots, U_r$ with U_i simple and with the annihilators of the U_i pairwise relatively prime and prime to (r). Let M be a faithful Λ module which is torsion free and finitely generated. Then there is epimorphism f: $M \to U$.

Proof. It suffices to show there exists an epimorphism $f_i: M \to U_i$. If so, take $f = (f_i): M \to \prod U_i = U$. If we localize at any $p \subset R$, at most one U_{ip} will be $\neq 0$. Therefore f is locally an epimorphism and thus is an epimorphism. Hence we can assume U is simple with annihilator p with $r \notin p$. Now $\Lambda_p = \Gamma_p$ is a maximal order. Since M is a faithful Λ module, $K \otimes M$ is a faithful A module. Therefore $\coprod_1^s KM = (KM)^s = A \oplus W$ for big enough s since A is semisimple. Thus by Theorem 5.27, $M_p^s \cong \Lambda_p \oplus N_p$ for some Λ module $N \subset W$ with $KN = W$. Hence, Λ_p is a direct summand of M_p^s. Now $U_p = U$ since all elements of $R - p$ act as units on U. So we have an epimorphism $\Lambda_p \to U$. In the sequence of maps $\coprod_1^s M_p \to \Lambda_p \to U \to 0$, one composite $M_p \to U$ must be non zero and, hence, onto since U is simple. Thus in the composite $M \to M_p \to U$ we have a non zero image. Again the simplicity of U implies the map is onto. Done.

Theorem 6.8 (Roiter). Let Λ be an R-order in a semisimple separable K algebra A where K is the quotient field of R and R is a Dedekind ring satisfying the Jordan-Zassenhaus Theorem. Let M, N_1, and N_2 be torsion free finitely generated Λ modules with $N_1 \vee N_2$ and M faithful. Then there exists an M' with $M' \sim M$ such that $M \oplus N_1 \approx M' \oplus N_2$.

Proof. There exists $r \neq 0$ in R and a maximal order Γ containing Λ with $r\Gamma \subset \Lambda$. Let $\underline{a} = (r)$ in lemma 6.6. Then we get an exact sequence $0 \to N_1 \to N_2 \to U \to 0$ where U satisfies the conditions of lemma 6.7. This lemma gives us an exact sequence $0 \to M' \to M \to U \to 0$. If $r \notin p$, then $U_p = 0$. Hence $M' \sim M$ by lemma 6.5. By the following lemma we have $N_1 \oplus M \approx M' \oplus N_2$.

Lemma 6.9 (Roiter). Let

$$0 \to A \to B \to W \to 0$$
$$0 \to C \to D \to W \to 0$$

be short exact sequences with B and D finitely generated and torsion free over Λ. Assume W is a torsion module and $W_y = 0$ for all y such that Λ_y is not a maximal order. Then $A \oplus D \approx B \oplus C$.

Proof. Form the pullback diagram

$$\begin{array}{ccccccccc}
 & & & & 0 & & 0 & & \\
 & & & & \downarrow & & \downarrow & & \\
 & & & & A & \xrightarrow{=} & A & & \\
 & & & & \downarrow & & \downarrow & & \\
0 & \to & C & \to & P & \to & B & \to & 0 \\
 & & {=}\downarrow & & \downarrow & & \downarrow j & & \\
0 & \to & C & \to & D & \xrightarrow{k} & W & \to & 0 \\
 & & & & \downarrow & & \downarrow & & \\
 & & & & 0 & & 0 & &
\end{array}$$

where $P = \{(b, d) | j(b) = k(d)\} \subset B \oplus D$. If the sequences $0 \to A \to P \to D \to 0$ and $0 \to C \to P \to B \to 0$ both split, then P would be isomorphic to both $A \oplus D$ and $C \oplus B$ which is what we want. If there were maps $f: D \to B$ and $g: B \to D$ such that $jf = k$ and $kg = j$ then the sequences would split for $D \xrightarrow{(f,1)} B \times D$ would have image contained in P and would split the sequence $0 \to A \to P \to D \to 0$. Thus we must find a map $f: D \to B$ which makes

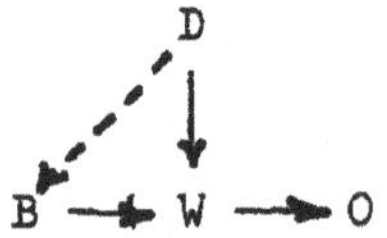

commute. This is possible locally. If Λ_y is not a maximal order we have $W_y = 0$. If Λ_y is a maximal order, D_y is projective. All modules here are finitely generated so Hom localizes. Since $\mathrm{Hom}_{\Lambda}(D, B) \to \mathrm{Hom}_{\Lambda}(D, W)$ is locally onto, it is onto and we are done.

Theorem 6.8 has the following corollary.

<u>Corollary</u> (Jacobinski). Let R be a Dedekind ring satisfying the Jordan-Zassenhaus Theorem. Let Λ be an R-order in a semisimple separable K-algebra A. Then $K_0(\Lambda)$ and $G_0(\Lambda)$ are finitely generated.

<u>Proof</u>. This was done for $G_0(\Lambda)$ in Chapter 2. The argument for $K_0(\Lambda)$ is given more generally in theorem 6.10 below.

Definition. Let Λ be an order and M a torsion free finitely generated Λ module. $\underline{D_M}$ is the abelian group with generators [N] for all N such that N is a direct summand of M^s for some s and with relations $[N_1] + [N_2] = [N_1 \oplus N_2]$.

Remark. It is easy to see that in D_M $[N_1] = [N_2]$ only if there exists an s with $M^s \oplus N_1$ isomorphic to $M^s \oplus N_2$. The argument is the same as that for $K_0(\Lambda) = D_\Lambda$.

Jacobinski generalizes corollary 6.9 as follows.

Theorem 6.10. If R satisfies the Jordan-Zassenhaus Theorem then D_M is finitely generated.

As always R is Dedekind and Λ is an order in a semisimple separable K-algebra.

Proof. Let $\Gamma \supset \Lambda$ be a maximal order. Let $r \neq 0$ in R with $r\Gamma \subset \Lambda$ and let p be a finite nonempty set of primes of R containing all $p \supset (r)$. Then $\Lambda_p = \Gamma_p$ is a maximal order if $p \notin P$. Now $\hat{M}_p$ the completion of M at p is a torsion free finitely generated $\hat{\Lambda}_p$ module. Define $D_M \to \prod_{p \in P} D_{\hat{M}_p}$ by sending [N] to $([\hat{N}_p])$. $D_{\hat{M}_p}$ is a finitely generated free abelian group by the Krull Schmidt Theorem. In fact, the only indecomposable modules that can occur in a direct summand of $\hat{M}_p^s$ are the ones that occur in $\hat{M}_p$. Hence, the image of D_M is finitely generated. Let $[N_1] - [N_2]$ be in the kernel. Then $[\hat{N}_{1_p}] - [\hat{N}_{2_p}] = 0$ in $D_{\hat{M}_p}$ for all $p \in P$. Thus, $\hat{N}_{1_p}$ is isomorphic to $\hat{N}_{2_p}$ for all $p \in P$ by the Krull Schmidt Theorem. Hence by lemma 6.5 $N_1 \sim N_2$. Let I be the annihilator in A of KM. Then $A_1 = A/I$ is still semi-simple and separable. $\Lambda_1 = \Lambda/I \cap \Lambda$ is an order in A_1, and M is a Λ_1 module. As a Λ_1 module M is faithful. D_M remains the same if computed over Λ or Λ_1. Hence, we can assume M is a faithful Λ module. Apply theorem 6.8 to obtain M' with $M' \sim M$ and $M \oplus N_1$ isomorphic to $M' \oplus N_2$. In D_M we have $[N_1] - [N_2] = [M'] - [M]$. By the Jordan-Zassenhaus hypothesis there are only finitely many such M' up to isomorphism. Thus the kernel is finite. Done.

<u>Remark</u>. This also shows that if $N_1 \sim N_2$ then $[N_1] - [N_2]$ has finite order since $[N_1] - [N_2]$ will be in the kernel of $D_M \rightarrow \prod_{p \in P} D_{\hat{M}_p}$. This can be strengthened as follows.

<u>Theorem 6.11</u>. Let R be a Dedekind ring satisfying the Jordan-Zassenhaus Theorem. Let M and N be torsion free finitely generated Λ modules with $M \sim N$. Then there exists an s such that M^s is isomorphic to N^s.

The converse is clear by completing and applying the Krull Schmidt Theorem.

<u>Proof</u>. We can assume M is faithful by the construction used in the proof of theorem 6.10. Then N is faithful since the annihilators of M and N are the same (we can embed N in M by Roiter's Lemma 3.1). By Theorem 6.8, we can find M' with $M \oplus M \approx M' \oplus N$.

Therefore $[N] \notin D_M$. By the remark preceding Theorem 6.11, $[M] - [N]$ has order $k < \infty$. Thus $[M^k] = [N^k]$ so $M^r \oplus M^k \approx M^r \oplus N^k$ for all large r. If we can show $(M^k)^t \approx (N^k)^t$ we can let $s = kt$. Therefore we replace M and N by M^k and N^k. We now have $M^t \oplus M \approx M^t \oplus N$ for all large t by the isomorphism just proved.

In other words it will suffice to prove the theorem under the additional hypothesis that $M^t \oplus M \approx M^t \oplus N$ for all $t \geq t_0$. This implies $M^t \oplus N^r = (M^t \oplus N) \oplus N^{r-1} = M \oplus (M^t \oplus N^{r-1}) = \ldots = M^t \oplus M^r$ for all $r \geq 1$, $t \geq t_0$.

Let $N_1, \ldots, N_h$ represent all isomorphism classes of modules $N_i \sim N$. Note $h < \infty$ by the Jordan-Zassenhaus Theorem. If for some n we have $M^n \oplus N_i \approx M^n \oplus M$, choose one such n and call it n_i. Choose $n_0 > n_i$ for $i = 1, \ldots, h$.

By Theorem 6.8, if $m \geq 1$, we have $N^m \oplus N \approx M^m \oplus N_{i_m}$ for some i_m. If $t \geq t_0$, then $M^{t+m+1} \approx M^t \oplus M^{m+1} \approx M^t \oplus N^{m+1} \approx M^t \oplus M^m \oplus N_{i_m}$. By the choice of n_0, we have $M^{n_0+1} \approx M^{n_0} \oplus N_{i_m}$. Therefore if we start with $m > n_0$, then $N^{m+1} = M^m \oplus N_{i_m} = M^{m-n_0} \oplus M^{n_0} \oplus N_{i_m} \approx M^{m+1}$.

As always R is a Dedekind ring and Λ an R-order in a semisimple separable K-algebra.

<u>Theorem 6.12</u>. (Jacobinski). Let P be a finite non empty set of primes such that if $p \notin P$ then Λ_p is a maximal order. If M and N

are finitely generated torsion free Λ modules such that $\hat{M}_p$ is a direct summand of $\hat{N}_p$ for all $p \in P$, then there exists M' such that $M' \sim M$ and M' is a direct summand of N.

Corollary 6.13. If M is locally a direct summand of N, then there exists M' such that $M' \sim M$ and M' is a direct summand of N.

Proof of theorem. Let $P = \{p_1, \ldots, p_n\}$. For each v with $1 \leq v \leq n$ we have $\hat{M}_{p_v} \xrightarrow{i_v} \hat{N}_{p_v} \xrightarrow{j_v} \hat{M}_{p_v}$ such that $j_v i_v$ is an isomorphism.

Now $\widehat{\mathrm{Hom}_{\Lambda_{p_v}}(M_{p_v}, N_{p_v})} = \hat{R}_{p_v} \otimes \mathrm{Hom}_{\Lambda_{p_v}}(M_p, N_p) = \mathrm{Hom}_{\hat{\Lambda}_{p_v}}(\hat{M}_{p_v}, \hat{N}_{p_v})$ since M and N are finitely presented and $\hat{R}_{p_v}$ is flat over R_{q_v}.

Hence, there exist $i_v' \in \mathrm{Hom}_{\Lambda_{p_v}}(M_{p_v}, N_{p_v})$ and $j_v' \in \mathrm{Hom}_{\Lambda_{p_v}}(N_{p_v}, M_{p_v})$ such that $i_v' \equiv i_v \bmod p_v$ and $j_v' \equiv j_v \bmod p_v$. Hence $j_v i_v' \equiv$ $\equiv ji \bmod p_v$, so by Nakayama's lemma, $j_v' i_v'$ is onto. But M_p has finite rank and is torsion free. Hence $j_v' i_v'$ is an isomorphism for all v. Now we can forget the completions. We have

$M_{p_v} \xrightarrow{i_v} N_{p_v} \xrightarrow{j_v} M_{p_v}$ with composition an isomorphism. Now $\mathrm{Hom}_{L_{p_v}}(M_{p_v}, N_{p_v}) = \mathrm{Hom}_L(M, N)_{p_v}$ since M and N are finitely presented. Say $i_v = i_v'/s_v$ and $j_v = j_v'/t_v$ with $s_v, t_v \in R - p_v$, $i_v' \in \mathrm{Hom}_L(M, N)$, and $j_v' \in \mathrm{Hom}_L(N, M)$. We can replace i_v and j_v

by i'_v and j'_v and the composition will still be an isomorphism. By the Chinese Remainder Theorem there are $i \in \mathrm{Hom}_L(M, N)$ and $j \in \mathrm{Hom}_L(N, M)$ such that $i \equiv i_v \bmod p_v$ and $j \equiv j_v \bmod p_v$ for all v. Then $ji: M_{p_v} \to M_{p_v}$ is onto for all v by Nakayama's Lemma and hence an isomorphism since M_{p_v} is torsion free finitely generated.

We claim $M' = j(N)$ will do. We need to show that $M \to j(N)$ is a split epimorphism and that $j(N) \sim M$. For $p \in P$, $j(N)_p \to M_p$ is an isomorphism. Hence M and j(N) have the same genus Lemma 6.5. If $p \in P$ then $N_p \to j(N)_p = M_p$ splits by construction. If $p \notin P$, then Λ_p is a maximal order and hence $j(N)_p$ is projective since it is finitely generated torsion free. Thus $N_p \to j(N)_p$ splits for all p. The proof of Theorem 6.12 and its corollary will be completed with the proof of the following lemma.

Lemma 6.14. If C and D are finitely generated modules over Λ and $f: C \to D$ is locally a split epimorphism then f is a split epimorphism.

Proof. f is an epimorphism since the coker of f is locally 0. We have the exact sequence

$$0 \to B \to C \xrightarrow{f} D \to 0.$$

A splitting will be a $k: D \to C$ with $fk = 1_D$. If the map $(1, f): \mathrm{Hom}_\Lambda(D, C) \to \mathrm{Hom}_\Lambda(D, D)$ is onto, there will be such a k. Now D is finitely generated and hence finitely presented. Thus

Hom_Λ localizes and hence, the coker of $Hom_\Lambda(D, C) \to Hom_\Lambda(D, D)$ is locally 0 and hence 0. So $Hom_\Lambda(D, C) \to Hom_\Lambda(D, D)$ is onto and a splitting exists.

Corollary 6.15. Let R satisfy the Jordan-Zassenhaus Theorem. Let M and N satisfy the hypothesis of Theorem 6.12. If every simple A module S which occurs in KN occurs strictly more times in KN than in KM, then M is a direct summand of N.

Proof. By Theorem 6.12 (which we assume applies), $N = M' \oplus P$ with $M' \sim M$. If P were faithful we could apply Theorem 6.8 to get some $P' \sim P$ with $M' \oplus P \cong M \oplus P'$ then $N \cong M \oplus P'$ and M would be a summand of N. Let $\underline{a}$ be the annihilator in A of KN. We pass to $\Lambda/\underline{a} \cap \Lambda = \Lambda_1$. All modules concerned are Λ_1-modules since $KM \cong KM' \subset KN$. The condition on simple submodules of KN implies KP is faithful and hence P is faithful.

Corollary 6.16. If $M \sim N$, then M is indecomposable if and only if N is.

Proof. Immediate from the theorem.

Definition. The genus of M, G_M, is $\{N | M \sim N\}$.

Addition of genera is well defined. A genus is called indecomposible if any one of its elements is (and therefore all are by corollary 6.16).

Theorem 6.17 (Jacobinsky). Let Λ be an R order in a semisimple separable K-algebra where R is a Dedekind ring satisfying the Jordan-Zassenhaus Theorem. Let M be a torsion free finitely generated Λ module. Then there are only finitely many indecomposible modules N such that N is a direct summand of M^s for some s.

Corollary 6.18. If Λ and R are as above, there are only finitely many indecomposible projectives.

Proof of the theorem. Let $\mathcal{P}$ be a finite nonempty set of prime ideals such that Λ_p is maximal if $p \notin \mathcal{P}$. We have an exact sequence $0 \to T \to D_M \to \prod_{\mathcal{P}} D_{\hat{M}_p}$ where $T = \{[N_1] - [N_2] \mid \hat{N}_{1_p} \approx \hat{N}_{2_p}$ for all $p \in \mathcal{P}\} = \{[N_1] - [N_2] \mid N_1 \sim N_2\}$. Let $\mathcal{U} = \{G_N \mid N$ is a direct summand of M^s for some $s\}$. $\mathcal{U}$ is a monoid with addition defined by $G_N + G_{N'} = G_{N \oplus N'}$.

Lemma 6.19. $\mathcal{U}$ is finitely generated. We first prove the theorem from the lemma. Let N be an indecomposible summand of M^s. By corollary 6.16, G_N is also indecomposible in $\mathcal{U}$ since $G_N = G_{N'} + G_{N''}$ implies $N \sim N' \oplus N''$. If S is a finite set of generators for $\mathcal{U}$, this shows that $G_N \in S$ so N belongs to one of a finite set of genera. Hence the rank of indecomposibles is bounded and the Jordan-Zassenhaus Theorem implies there are only a finite number of indecomposibles.

Proof of Lemma 6.19. We map $\mathcal{U}$ to D_M/T by $G_N \rightsquigarrow [N]$ mod T. This is well defined, additive and one-one. In fact $[N] = [N']$ mod T implies that $[N] - [N'] =]M] - [N_2]$ where $N_1 \sim N_2$. Thus for some s, $M^s \oplus N \oplus N_2 \approx M^s \oplus N' \oplus N_1$ so $N \sim N'$. However, it is not true that a submonoid of a finitely generated abelian group is finitely generated. For example, the submonoid of $Z \times Z$ given by $\{(m, n) | (m, n) = (0, 0) \text{ or } m > 0 \text{ and } n > 0\}$ is not finitely generated.

We need to characterize the image more precisely. Now $D_{\hat{M}_p}$ is free abelian on $[P_1], \ldots, [P_s]$ where $P_1, \ldots, P_s$ are the indecomposibles occuring in a direct sum decomposition $\hat{M}_p = \coprod_1^s P_i^{n_i}$. We define an isomorphism of $D_{\hat{M}_p}$ with $Z \oplus \ldots \oplus Z$ (s times) by $[N] \rightsquigarrow (f_1([N]), \ldots, f_s([N]))$ where $f_i([N])$ = the number of times P_i occurs in the decomposition of N into a direct sum of indecomposibles. We extend the map f_i to D_M by $f_{ip_j}: D_M \to \prod D_{\hat{M}_p} \to D_{\hat{M}_{p_j}} \xrightarrow{f_i} Z$. The f_{ip_j} all anihilate T since T is a finite group. Hence we have functions f_{ip_j} defined on D_M/T. Identifying G with its image in D_M/T we claim that if $x \in D_M/T$, then $x \in G$ if and only if all $f_{ip_j}(x) \geq 0$. If $x \in G$, then all $f_{ip_j}(x) \geq 0$ clearly since $x = [N]$ for some N. Suppose all $f_{ip_j}(x) \geq 0$. Say $x = [N] - [N']$. Now $x \rightsquigarrow y \in D_{\hat{M}_p}$ where $y = [\hat{N}_p] - [\hat{N}'_p]$. In $D_{\hat{M}_p}$ there does exist a module P_p such that $y = [P_p]$ namely $P_p = \coprod P_i^{f_i(y)}$ By the Krull-Schmidt Theorem,

$\hat{N}_p$ is isomorphic to $\widehat{N'_p} \oplus P_p$ for all $p \in P$. Hence by Theorem 6.12, there exists an $N'' \sim N'$ and a Q such that $N = N'' \oplus Q$. Now $x = [N] - [N'] = [N''] - [N'] + [Q] = [Q] \bmod T$ since $N'' \sim N'$ implies $[N''] - [N'] \in T$. Hence $x = [G_Q] \in \mathcal{G}$ as claimed.

Lemma 6.19 now follows from the following generalization of a classical result.

Theorem 6.20. (Gordon's Lemma). Let A be a finitely generated commutative monoid and let $f_i: A \to Z$ $i = 1, \ldots, n$ be a finite collection of homomorphisms. Then $G = \{x \in A | f_i(x) \geq 0 \text{ for all } i\}$ is a finitely generated monoid.

We note that a finitely generated abelian group A is also finitely generated as a monoid. If a_i generate A as a group, then all $\pm a_i$ generate A as a monoid.

Remark. Let α be an irrational real number. Define $f: Z \oplus Z \to \mathbb{R}$ by $f((m, n)) = m - n\alpha$. Then $G = \{x | f(x) \geq 0\}$ is not a finitely generated monoid. This shows that we cannot replace Z by $\mathbb{R}$ or even by $Z + Z\alpha \subset \mathbb{R}$ in this theorem. Since $m - n\alpha \geq 0$ if and only if $m - nr \geq 0$ for all rational $r < \alpha$, the same G can be defined by an infinite number of maps into Z. Thus the restriction to a finite number of f_i is essential. The example $G = \{(x, y) \in Z \times Z | x > 0 \text{ and } y > 0\}$ shows that we cannot replace the condition $f_i(x) \geq 0$ by $f_i(x) > 0$.

Proof. We use induction on n, the number of f_i. If $G_1 = \{x \in A | f_i(x) \geq 0 \text{ for } 1 \leq i \leq n - 1\}$ then G_1 is finitely generated by the induction hypothesis and $G = \{x \in G_1 | f_n(x) \geqslant 0\}$. Therefore it will suffice to do the case $n = 1$. We write $f = f_1$.

Let $a_1, \ldots, a_r$ generate A. Any element of A has the form $a = x_1a_1 + \ldots + x_ra_r$ where all $x_i \in Z$, $x_i \geqslant 0$. Let $p_i = f(a_i)$. Then $f(a) \geqslant 0$ if and only if the x_i satisfy $\sum x_i f(a_i) = \sum x_i p_i \geqslant 0$ or $x_0 \geqslant 0$ where we set $x_0 = \sum p_i x_i$.

Let H be the set of $(x_0, \ldots, x_r) \in Z^{r+1}$ satisfying the conditions $x_i \geqslant 0$ for $0 \leqq i \leqq r$ and $x_0 = \sum p_i x_i$. Then H is a monoid under addition and clearly $G = \{\sum_1^n x_i a_i | \exists x_0$ with $(x_0, \ldots, x_n) \in H\}$. Therefore the map φ: $H \to G$ by $\varphi(x_0, \ldots, x_n) = x_1a_1 + \ldots + x_ra_r$ is a monoid homomorphism and is onto. Thus it will suffice to show that H is a finitely generated monoid.

We relabel $x_0, x_1, \ldots, x_n$ as $u_1, \ldots, u_a, v_1, \ldots, v_b, w_1, \ldots, w_c$ in some order so that the equation $x_0 = \sum p_i x_i$ takes the form $r_1u_1 + \ldots + r_au_a = s_1v_1 + \ldots + s_bv_b$ where all $r_i > 0$ and all $s_j > 0$.

Define elements $h_i, h_{ij} \in H$ by letting h_i have $w_i = 1$ and all other coordinates 0 and by letting h_{ij} have $u_i = s_j$, $v_j = r_i$ and all other coordinates zero. If $h = (x_0, \ldots, x_n) \in H$, let $|h| = \sum_0^r x_i$. Then $|h + h'| = |h| + |h'|$. If h has $w_i \neq 0$ for some i, we can write $h = h_i + h'$ where $|h'| < |h|$. If all $w_i = 0$ but some $u_i > N$ (an integer to be determined) then $N \cdot \min(r_i) < \sum r_i u_i = \sum s_j v_j \leq (\sum s_j) \max(v_j)$ so some $v_j > \alpha N$ where $\alpha = (\sum s_j)^{-1}(\min r_i)$. Choose N so that N and αN are greater than all r_i, s_j (and also > 1). Then we will have $u_i > s_j$, $v_j > r_i$ so we can write $h = h_{ij} + h'$ where $|h'| < |h|$.

By induction on $|h|$ we see that H is generated by the h_i, h_{ij} and those $h \in H$ with all $w_i = 0$ and all $u_i \leq N$. But there are only a finite number of such h since $u_i \leq N$ for all i implies $v_j \leq \sum s_j v_j = \sum r_i u_i \leq (\sum r_i)N$.

Theorem 6.21 (Roiter). Let Λ be an order over a Dedekind ring R in a separable semisimple algebra over K, the quotient field of R. Let R satisfy the Jordan-Zassenhaus Theorem and assume that R/p is finite for each nonzero prime p of R. Then there exists an integer N such that every genus of Λ-modules contains $\leq$ N isomorphism classes.

Note that the hypothesis on R/p holds if R is the ring of integers of an algebraic number field or a finite extension of k[t] with finite k.

Proof. Let G be a genus and let $A = A_1, A_2, \ldots, A_n$ be representatives of all the isomorphism classes of G. Let $r \neq 0$ be an element of R such that $r\Gamma \subset \Lambda$ where $\Gamma \supset \Lambda$ is a maximal order. By lemma 6.6, we can embed each A_i in A such that A/A_i is a direct sum of simple modules whose orders are relatively prime and prime to r, and such that if $i \neq j$ the orders of A/A_i and A/A_j are relatively prime. Let $B = \bigcap A_i$ in A and let $U = A/B$. Then U is a direct sum of simple modules of relatively prime orders and orders relatively prime to r. In fact $A/B \to \prod A/A_j$ is clearly injective and is an epimorphism since this is true locally. By Lemma 6.7 there is an epimorphism $\Lambda \to U$. Since Λ is projective there exists a map $\Lambda \to A$ making the diagram

$$\begin{array}{ccccccccc} & & & & & & \Lambda & & \\ & & & & & \swarrow & \downarrow & & \\ 0 & \to & B & \to & A & \to & U & \to & 0 \\ & & & & & & \downarrow & & \\ & & & & & & 0 & & \end{array}$$

commute.

Let C be the image of $\Lambda \to A$. Then $\operatorname{rk} C \leq \operatorname{rk} \Lambda$ and $B + C = A$. We want such a C which is pure in A (with respect to R). Let $D = (KC) \cap A$ in KA. Then $\operatorname{rk} D = \operatorname{rk} C$, $D \supset C$, $B + D = A$, A/D is torsion free, and $\operatorname{rk} D \leq \operatorname{rk} \Lambda$. Since $A_i \supset B$, we have $A_i + D \supset B + D = A$. Therefore $A_i + D = A$ so $A_i \to A/D$ is onto. Consider the following commutative diagram with exact rows and columns:

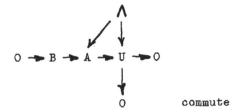

$$\begin{array}{ccccccccc}
0 & \longrightarrow & A_i \cap D & \longrightarrow & A_i & \longrightarrow & A/D & \longrightarrow & 0 \\
 & & \uparrow{\scriptstyle f_i} & & \uparrow{\scriptstyle g_i} & & \uparrow{\scriptstyle 1} & & \\
0 & \longrightarrow & B \cap D & \xrightarrow{j} & B & \xrightarrow{e} & A/D & \longrightarrow & 0 \\
 & & \uparrow & & \uparrow & & \uparrow & & \\
 & & 0 & & 0 & & 0 & &
\end{array}$$

We can recover A_i from f_i by forming the pushout. Hence we need to calculate the number of possible pushouts.

If $h: B \to A_i \cap D$, we can get a new commutative diagram by replacing g_i by $g_i + h$ and f_i by $f_i + hj$. Thus f_i and $f_i + hj$ yield the same pushout. Therefore the number of possible pushouts (up to isomorphism) is $\leq$ the order of the cokernel of

$$(j, 1): \mathrm{Hom}_\Lambda(B, A_i \cap D) \to \mathrm{Hom}_\Lambda(B \cap D, A_i \cap D).$$

We claim the image of $(j, 1)$ contains $r\,\mathrm{Hom}(B \cap D, A_i \cap D)$. Since A/D is torsion free, we can apply lemma 6.1 to the diagram

$$\begin{array}{ccccc}
 & & A/D & & \\
 & & \downarrow{\scriptstyle 1} & & \\
B & \xrightarrow[e]{} & A/D & \longrightarrow & 0
\end{array}$$

and get $m: A/D \to B$ with $em = r1_{A/D}$. Now $e(r1_B - me) = 0$ so we can write $r1_B - me = jk$ where $k: B \to B \cap D$. Now $jkj = rj - mej = rj$ but j is a monomorphism so $kj = r1_{B \cap D}$. Therefore, if $f: B \cap D \to A_i \cap D$, then $fk: B \to A_i \cap D$ and $(fk)j = rf$.

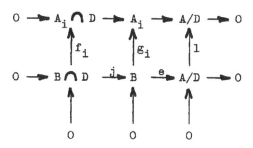

This proves our assertion. It follows from this that the number of possible pushouts is less than or equal to the order of $\mathrm{Hom}_\Lambda(B \cap D, A_i \cap D)/r\,\mathrm{Hom}_\Lambda(B \cap D, A_i \cap D)$. This order is finite since $R/(r)$ is finite and $\mathrm{Hom}_\Lambda(B \cap D, A_i \cap D)$ is finitely generated over R. Now $B \cap D \subset D$ and $A_i \cap D \subset D$. Hence $\mathrm{rk}(B \cap D)$ and $\mathrm{rk}(E)$ are bounded by $\mathrm{rk}(D) \leq \mathrm{rk}\,\Lambda$. Hence by the Jordan-Zassenhaus Theorem, there are only a finite number N_1 of possible pairs of $B \cap D$ and $A_i \cap D$. Let N_2 be the maximum of the orders of $\mathrm{Hom}(B \cap D, A_i \cap D)/r\,\mathrm{Hom}(B \cap D, A_i \cap D)$ over all such pairs. Then the number of possible A_i is less than or equal to $N_1 N_2$ for every genus.

Done.

Chapter 7: K_0 of a Maximal Order

In this chapter we will compute K_0 for maximal orders in algebras over global fields. Let R be a Dedekind ring with quotient field K, A a semisimple separable K-algebra and Λ a maximal R-order in A. Then $A = A_1 \times \dots \times A_r$ with A_i simple and $\Lambda = \Lambda_1 \times \dots \times \Lambda_r$. Let A_i have center K_i. Then Λ_i is a maximal R_i-order in A_i where R_i is the integral closure of R in K_i. Therefore $K_0(\Lambda) = \prod K_0(\Lambda_i)$. We are thus reduced to looking at Λ_i so it will suffice to consider the case where A is central simple over K. Let S be a simple Λ module. Then there exists a prime ideal $p \subset R$ such that $pS = 0$. By Theorem 5.16 there is a unique 2 sided prime P of Λ containing $p\Lambda$ and $p\Lambda = P^e$ some $e > 0$. Then Λ/P is a simple algebra over R/p. Since S is simple, $P^eS = 0$ implies PS = 0 so S is a Λ/P-module. Since Λ/P is simple, there is exactly one simple S with pS = 0 for any $p \subset R$. Hence the simple Λ modules are in 1-1 correspondence with the non-zero prime ideals of R. Let I_R be the group of fractional ideals of R. This is free abelian on the non-zero primes p of R. We map $K_0(\Lambda) \to K_0(A)$ by $[M] \rightsquigarrow [K \otimes_R M]$. This is clearly onto. For any finitely generated torsion Λ module C we have $\mathrm{pd}_\Lambda C \leq 1$ by Lemma 5.13 so there is a projective resolution $0 \to P' \to P \to C \to 0$. By Schanuel's lemma, $[P] - [P'] \in K_0(\Lambda)$ is independent of the choice of the resolution. Hence if we let T_Λ be the category of finitely generated torsion modules over Λ, there is a well defined map $K_0(T_\Lambda) \to K_0(\Lambda)$ sending [C] to [P] - [P']. The sequence $K_0(T_\Lambda) \to K_0(\Lambda) \to K_0(A) \to 0$ is exact. In fact, given $C \in T_\Lambda$ we have $K \otimes_R C = 0$ so $K \otimes_R P \approx K \otimes_R P'$. Conversely if $[P] - [P'] \mapsto 0$ in $K_0(A)$, then $K \otimes_R P \approx K \otimes_R P'$. By Theorem 5.27, $P \sim P'$ so by Roiter's lemma, we can find $0 \to P' \to P \to C \to 0$ with C a torsion module. Since A is simple, $K_0(A)$ is isomorphic to Z. Also $K_0(T_\Lambda)$ is free on the simple Λ-modules. Hence we can define an isomorphism $K_0(T_\Lambda) \approx I_R$ by $[S] \rightsquigarrow p$ where p is the unique non zero prime with pS = 0 above. Let $C \in T_\Lambda$. We want to describe the image of [C] in I_R in a more intrinsic way. Let O(C), the order ideal of C, be the ideal of R defined by $O(C) = \prod p_i$ where there exists a series of R submodules of C, $C = C_n \supset C_{n-1} \supset \dots \supset C_0$ with C_i/C_{i-1} isomorphic to R/p_i. This is well defined by the Jordan-Hölder theorem and depends only on the R-module structure of C. If $0 \to C' \to C \to C'' \to 0$ is exact, then we can construct a composition series for C (over R) from such series for C and C'. This shows that $O(C) = O(C')O(C'')$. Hence we have $O\colon K_0(T_\Lambda) \to I_R$. This is not the above isomorphism. If S is simple and $p \subset R$ is the unique prime with $pS = 0$, then S is an R/p module. Since R/p is a field, $O(S) = p^{\dim_{R/p} S}$ but $\dim_{R/p} S$ is not necessarily 1. For example if $\Lambda = M_n(R)$, the $n \times n$ matrices over R, then a simple module S with $pS = 0$ has $\dim_{R/p} S = n$. We want to show this also happens in the general case provided R/p is finite.

Theorem 7.1. Let R be a Dedekind ring with quotient field K. Let Λ be a maximal R-order in a central simple K-algebra A. If $\dim_K A = n^2$, R/p is finite, and S is a simple Λ module with $pS = 0$, then $\dim_{R/p} S = n$.

Proof. 1) Localize at p. Then Λ_p is a maximal order over R_p, $\Lambda/p_p\Lambda_p = \Lambda/p\Lambda$, and $R_p/p_pR_p = R/p$. Since S is a simple $\Lambda/p\Lambda$-module, nothing changes and so we can assume R is local and hence a discrete valuation ring.

2) Next we show that we can assume R is complete. Let $\hat{R}$ be the completion of R and $\hat{K}$ be the quotient field of $\hat{R}$. $\hat{A} = \hat{K} \otimes_K A \supset \hat{\Lambda} = \hat{R} \otimes_R \Lambda$. Then $\hat{\Lambda}/p\hat{\Lambda} = \Lambda/p\Lambda$ and $\hat{R}/p\hat{R} = R/p$.

We need to show that $\hat{\Lambda}$ is a maximal order. The remaining conditions are clearly preserved by completion.

Lemma 7.2. Let R be a discrete valuation ring with quotient field K. Let Λ be a maximal R-order in a semisimple, separable K-algebra. Then the completion $\hat{\Lambda}$ is a maximal order over $\hat{R}$ in $\hat{A} = \hat{K} \otimes_R A$.

Proof. Let $\Gamma \supset \hat{\Lambda}$ be a maximal $\hat{R}$-order in $\hat{A}$. We want to show $\Gamma = \hat{\Lambda}$. This is accomplished by

Lemma 7.3. Let R be a discrete valuation ring, $\hat{R}$ its completion, K be the quotient field of R, $\hat{K}$ the quotient field of $\hat{R}$. Let V be a finite dimensional vector space over K, $\hat{V} = \hat{K} \otimes_K V$, and let M be a finitely generated $\hat{R}$ submodule of $\hat{V}$. Then

(1) $N = M \cap V$ is a finitely generated R module, and

(2) If $\mathrm{rk}\, M = \mathrm{rk}\, \hat{V}$, then $\hat{R}N = M$.

Remark. This proves lemma 7.2 since $\Gamma \cap A$ is an order by (1) and $\Gamma \cap A \supset \Lambda$. Since Λ is maximal, $\Lambda = \Gamma \cap A$ and (2) implies $\Gamma = \hat{R}\Lambda = \hat{\Lambda}$.

Proof. Let $e_1, \ldots, e_n$ be a basis for V over K (and hence also for $\hat{V}$ over $\hat{K}$). Let $m_1, \ldots, m_r$ be a basis for M over $\hat{R}$. Then $m_i = \sum a_{ij} e_j$ with $a_{ij} \in \hat{K}$. If $m \in M$, then $m = \sum g_i m_i$ with $g_i \in \hat{R}$ so $\mathrm{ord}\, g_i \geq 0$. Thus $m = \sum g_i \alpha_{ij} e_j$ and $\mathrm{ord}(\sum_i g_i \alpha_{ij}) \geq \min(\mathrm{ord}\, \alpha_{ij}) = s$. This shows that if $m \in M$ and $m = \sum t_j e_j$, then $\mathrm{ord}\, t \geq s$ for all j. Let $n \in N = M \cap V$. Then

$n = \sum t_j e_j$ with $t_j \in K$ and we have just shown that $\operatorname{ord}(t_j) \geq s$. Hence $N \subset \sum_j R\pi^s e_j$ which is a finitely generated R module. (Here π is any element of R with $R\pi = p$). Hence N is a finitely generated module since R is noetherian.

We now prove part (2). Here we have $r = n$ so the m_i form a basis for $\hat{V}$ over $\hat{K}$. Hence (a_{ij}) is an invertible matrix with inverse (b_{ij}) say. Thus $e_j = \sum b_{jk} m_k$. We want to show that N is dense in M. Let $m \in M$. Then $m = \sum g_i m_i = \sum g_i a_{ij} e_j = \sum t_j e_j$ with $t_j \in \hat{K}$. We approximate the t_j by $t_j' \in K$. Then $x = \sum t_j' e_j \in V$ since $t_j' \in K$. But $\sum t_j' e_j = \sum g_i' a_{ij} e_j = \sum g_i' m_i$ and the $g_i' = \sum b_{ki} t_k'$ are close to the $g_i = \sum b_{ki} t_k$. Hence if the t_j' are picked close enough all the ord g_i' will be positive since ord $g_i \geq 0$. Hence all the $g_i' \in \hat{R}$. Thus $x = \sum g_i' m_i \in M$. Hence $x \in M \cap V = N$. Since x can be made arbitrarily close to m by taking the t_i' close to t_i we can approximate $m \in M$ by elements of N and so N is dense in M.

Hence in the proof of Theorem 7.1 we can assume R is complete local discrete valuation ring. We now proceed with the proof of Theorem 7.1. By Wedderburn theory we know that $A = M_r(D)$, the $r \times r$ matrices over a division ring D.

Lemma 7.4. If Γ is a maximal order of D, then $M_r(\Gamma)$ is a maximal order of A.

Proof. Let Λ be a maximal order containing $M_r(\Gamma)$ and let $\Lambda_0 =$ $\{d \in D \mid \begin{pmatrix} d & 0 & 0 \\ 0 & & \\ 0 & & 0 \end{pmatrix} \in \Lambda\}$. Then $\Lambda_0 \subset D$ and is finitely generated over R. Also $\Lambda_0 \supset \Gamma$, $\Lambda_0 \Lambda_0 \subset \Lambda_0$ and $1 \in \Lambda_0$. Hence Λ_0 is an order of D containing Γ so $\Lambda_0 = \Gamma$. Let $(t_{ij}) \in \Lambda$ with $t_{ij} \in D$. Then

$$\begin{pmatrix} 0 \ldots 01 \ldots 0 \\ \bigcirc \end{pmatrix} \begin{pmatrix} t_{ij} \end{pmatrix} \begin{pmatrix} 0 & \\ \vdots & 0 \\ 0 & \end{pmatrix} = \begin{pmatrix} t_{km} & 0 \ldots 0 \\ 0 & \\ \vdots & \bigcirc \\ 0 & \end{pmatrix} \in \Lambda_0$$

where the first matrix has a 1 in the k-th column and 1st row and the last has a 1 in the 1st column m-th row. Thus $t_{ij} \in \Gamma$ for all i and j. Hence $M_r(\Gamma) = \Lambda$. This proves Lemma 7.4.

We now observe that it will suffice to prove Theorem 6.1 for a single maximal order Λ in A. If Λ' is another and $P = \Lambda'\Lambda$, then Th. 5.15 shows that the functor $S \mapsto P \otimes_\Lambda S$ gives an equivalence between the categories of Λ modules and Λ - modules. Since simplicity can be defined in terms of the category, simple modules will correspond. Since R is local, $P \approx \Lambda$ as a right Λ module by Theorem 5.27 so $P \otimes_\Lambda S \approx S$ as an R-module and our assertion follows immediately.

We can now assume $\Lambda = m_r(\Gamma)$. Now $M_r(\Gamma)/pM_r(\Gamma) =$ $M_r(\Gamma/p\Gamma)$ and factoring out the radical gives $M_r(\Gamma/\underline{P})$. If S' is a simple $\Gamma/\underline{P}$ module, then $S = \{ \begin{pmatrix} s_1 \\ \vdots \\ s_r \end{pmatrix} | s_i \in S' \}$ is a simple $M_r(\Gamma/P)$ module and $\dim_{R/p} S = r \dim_{R/p} S'$. If the theorem is true for Γ, then $\dim_{R/p} S' = n/r$, so $\dim_{R/p} S = n$. Hence, it is enough to prove the theorem in the case $A = D$. That is, we can assume A is a division ring and R is a complete DVR. This implies that $\Lambda/\underline{P}$ is a division ring. To see this note that $\Lambda/\underline{P}$ is simple and so $\Lambda/\underline{P} = M_S$ (division ring) by Wedderburn theory. We want to show $S = 1$. If not there is an idempotent $e \in \Lambda/\underline{P}$ with $e \neq 0, 1$. Since R is complete, we can lift e to an idempotent $e' \in \Lambda$. Since $e \neq 0, 1$ we have $e' \neq 0, 1$ but this is impossible since $e' \in A$, a division ring. Hence $\Lambda/\underline{P}$ is a division ring and is a finite module over R/p. Since R/p is finite by hypothesis, so is $\Lambda/\underline{P}$ and so $\Lambda/\underline{P}$ is a field (by Wedderburn's theorem). Now $p\Lambda$ is a 2 sided ideal of Λ and $p\Lambda = \underline{P}^e$ for some e. Since R is local, $\underline{P}$ is principal. Let $\underline{P} = \Lambda a$.

We claim that $\underline{P}^r = \Lambda a^r$. Assume this for some r. Since $\underline{P}$ is 2-sided, $\Lambda a = \underline{P} = \underline{P}\Lambda = \Lambda a \Lambda$ so $P^{r+1} = \Lambda a \Lambda a^r = \Lambda a a^r = \Lambda a^{r+1}$. Mapping $\Lambda/P \rightarrow \underline{P}^r/\underline{P}^{r+1}$ by $1 \rightsquigarrow a^r$ gives an isomorphism. It is clearly onto. Since $ta^r = 0$ in P^r/P^{r+1} if and only if $ta^r \in \underline{P}^{r+1}$, we have $ta^r = ua^{r+1}$ for such t. But we can cancel the a^r since A is a division ring. Therefore $t = ua \in \Lambda a = \underline{P}$. We filter $\Lambda/p\Lambda$ as $\Lambda \supset P \supset P^2 \supset \ldots \supset P^e = p\Lambda$. Thus $\dim_{R/p} \Lambda/p\Lambda = ef$ where $f = \dim_{R/p} \Lambda/\underline{P} = \dim_{R/p} S$.

We want to show $f = n$. We know that $n^2 = \dim_{R/p} \Lambda / p\Lambda = ef$. Hence it is enough to show that 1) $e \leq n$ and 2) $f \leq n$. For 1), let $p = R\Pi$. Since $\underline{P} = \Lambda a$ and $p\Lambda = \underline{P}^e = \Lambda\Pi = \Lambda a^e$ we have $\Pi = ua^e$ and $a^e = v\Pi$ where u and v are units of Λ. Now $K(a) \subset A$ is a field since $K(a)$ is commutative and contained in a division ring. Also $[K(a): K] \leq n$ since every element of A satisfies an equation of degree $\leq n$ over K. We claim that the ramification index of $K(a)/K$ is greater than or equal to e. But (ramification index $K(a)/K$) $\times$ (residue class degree) = $[K(a): K]$. Note there is only one prime over p since R is complete. Also $R[u] \subset \Lambda$ which is a finitely generated R module. Therefore, u is integral over R. Similarly v is integral over R. Let R' be the integral closure of R in K(a). Then $u, v \in R'$ with $uv = 1$. Since $\Pi = ua^e$ and $a^e = v\Pi$, we have $\operatorname{ord}\Pi = e \operatorname{ord} a$ in R'. Hence the ramification index of K(a) over K is $\geq e$ and $e \leq n$ as claimed.

To prove $f \leq n$ we need the fact that R/p is finite. This implies that $\Lambda/\underline{P}$ is a finite separable field extension of R/p and hence $\Lambda/\underline{P} = (R/p)(b)$. (Note: this would also be true if R/p were algebraically closed.) There exists a $g \in \Lambda$ with $b \equiv g \bmod \underline{P}$ and such that g satisfies a monic equation of degree n with coefficients in R, i.e., its minimal equation. Therefore, b satisfies the equation obtained from this by reduction mod p and so

$f = [\Lambda/\underline{P}: R/p] = \deg b \leq n$. Hence $e, f \leq n$ and $ef = n^2$ so $f = n$ as desired. This completes the proof of Theorem 7.1.

Hence $O(S) = p^n$. Thus if C is a torsion Λ module then $O(C)$ is an n-th power in I_R. (Use a composition series for C.)

<u>Definition</u>. Let C be a finitely generated torsion Λ module. Then the <u>reduced order ideal</u> of C is defined to be $\mathcal{O}_r(C) = O(C)^{1/n}$.

If S is simple and $pS = 0$, then $\mathcal{O}_r(C) = p$. Therefore $\mathcal{O}_r: K_0(T_\Lambda) \to I_R$ is the desired isomorphism. The exact sequence

$$K_0(T_\Lambda) \to K_0(\Lambda) \to K_0(A) \to 0$$

reduces to $I_R \to K_0(\Lambda) \to Z \to 0$ since $K_0(A)$ is isomorphic to Z and $K_0(T_\Lambda)$ is isomorphic to I_R. We must now determine the kernel of $I_R \to K_0(\Lambda)$.

<u>Lemma 7.5</u>. The kernel of the map $K_0(T_\Lambda) \to K_0(\Lambda)$ is generated by all [C] such that there exists a finitely generated free Λ-module F and an exact sequence $0 \to F \to F \to C \to 0$.

<u>Proof</u>. Clearly all such [C] belong to the kernel. Let $x \in K_0(T_\Lambda)$ go to 0 in $K_0(\Lambda)$. Write $x = [C] - [D]$ for some $C, D \in T_\Lambda$. Let $0 \to P \to F \to D \to 0$ be exact with F finitely generated and free. Since D is torsion, there exists an $s \neq 0$ in R such that $sD = 0$. Hence $sF \subset P$ and we have an exact sequence

$$0 \to D' \to F/sF \to D \to 0 .$$

Thus $x = [C \oplus D'] - [D \oplus D']$ and $[D \oplus D'] = [F/sF]$. Therefore, in the expression $x = [C] - [D]$ for any $x \in K_0(T_\Lambda)$ we can assume D has the form F/sF. Therefore $0 \to F \xrightarrow{s} F \to D \to 0$ so $[D]$ is one of the specified elements. If x goes to 0 in $K_0(\Lambda)$ then so does $[C]$ since $[D]$ clearly does.

Thus we are reduced to showing that if $[C]$ goes to 0 in $K_0(\Lambda)$ then there is a finitely generated free F and an exact sequence $0 \to F \to F \to C \to 0$. Choose an exact sequence $0 \to P \to F \to C \to 0$ with F free. Then $[C] = [F] - [P]$ in $K_0(\Lambda)$ but $[C] = 0$ so $[F] - [P] = 0$. Therefore, there exists F' such that $F' \oplus P$ is isomorphic to $F' \oplus F$. Hence there is an exact sequence

$$0 \to F \oplus F' \to F \oplus F' \to C \to 0$$

as desired.

Let $[C] \in \mathrm{Ker}(K_0(T_\Lambda) \to K_0(\Lambda))$ and let $0 \to F \xrightarrow{f} F \to C \to 0$ be an exact sequence as in the lemma. Then $f \in \mathrm{End}(F) = M_s(\Lambda)$ where F is free on s generators. Clearly f is a monomorphism if and only if f is a unit in $M_s(A) \supset M_s(\Lambda)$. Now, the order ideal of $M_s(\Lambda)/fM_s(\Lambda)$ is $(N(f))$ where $N(f)$ is the usual norm of f. It is sufficient to check this locally. Assume R local so Λ is a free R-module. Let (e_i) be a base for $M_s(\Lambda)$ over R and so for $M_s(A)$ over K. Let $fe_i = \sum a_{ij}e_j$. Then $N(f) = \det(a_{ij})$.

Since R has been localized, (a_{ij}) is equivalent to a diagonal matrix diag$(d_1, \ldots, d_t)$. Thus $M_i(\Lambda)/fM_s(\Lambda) \approx \coprod R/d_i)$ as an R-module so its order ideal is $(d_1, \ldots, d_t)$. Now $M_s(\Lambda) = F \oplus \ldots \oplus F$ (s times). Therefore, $M_s(\Lambda)/fM_s(\Lambda) = \coprod_1^s F/fF$ and so

$$(N(f)) = O(\coprod_1^s F/fF) = O(\coprod_1^s C) = O(C)^s = \mathcal{O}_r(C)^{ns}.$$

If $n(f)$ is the reduced norm of f, then $N(f) = n(f)^{ns}$ since $\dim_R M_s(\Lambda) = n^2s^2$. Hence $\mathcal{O}_r(C) = n(f)$. Thus we have an exact sequence

$$0 \to X \to I_R \to K_0(\Lambda) \to K_0(A) \to 0$$

where X is generated by all $(n(f))$ where $f \in M_s(\Lambda)$ and f is a unit of $M_s(A)$.

Let a be a unit in $M_s(A)$. Then there exists an $r \neq 0$ in R such that $ra \in M_n(\Lambda)$. Now $r \cdot I \in M_s(\Lambda)$ and $r \cdot I$ is a unit in $M_s(A)$. Thus $n(ra)$ and $n(r)$ are in X. Since X is a group, $(n(a) = n(ra)/n(r) \in X$. Thus, X is generated by all $(n(f))$ where f is a unit in $M_s(A)$ as claimed.

Let $M_s(A)^* = GL_s(A)$ be the group of units of $M_s(A)$. Define $GL_s(A) \to I_R$ by $a \mapsto (n(a))$ and let Hs be the image of this. Then we have an exact sequence

$$\sum Hs \to I_R \to K_0(\Lambda) \to K_0(A) \to 0 .$$

Later we will show this is a part of an exact sequence of K-theory. We must now determine the Hs. Assume that K is a global field, i.e., an algebraic number field or function field of dimension 1 over a finite field

Our problem is to find the image of $n: M_s(A)^* \to K^*$. Since $M_s(A)$ is a central simple algebra over K, it will suffice to determine the image of $n: B^* \to K^*$ for any central simple K-algebra B. If y is a prime (i.e., a valuation) of K, let $\hat{K}_y$ and $\hat{B}_y = \hat{K}_y \otimes_K B$ denote the completions of K and B at y.

<u>Definition</u>. We say that y is unramified in B if and only if $\hat{B}_y = M_n(\hat{K}_y)$, a matrix algebra over $\hat{K}_y$.

In particular, if y is a real archimedian prime, then $\hat{K}_y = \mathbb{R}$ and $\hat{B}_y = M_n(\mathbb{R})$ or $M_m(\mathbb{H})$ where $\mathbb{H}$ denotes the quaternions. Such a y is ramified if and only if $B \approx M_m(\mathbb{H})$. The following classical result answers our question.

<u>Theorem 7.6</u>. (Hasse-Schilling norm theorem). Let K be a global field and B a central simple K-algebra. Then the image of $n: B^* \to K^*$ consists of those $a \in K^*$ which are positive at each real archimedian prime which is ramified in B.

<u>Corollary 7.7</u>. If char $K \neq 0$, $n: B^* \to K^*$ is onto.

To say that a is positive at y means that a corresponds to a positive real number under the unique isomorphism $\hat{K}_y \approx R$.

It is easy to see that the condition is necessary. We have a commutative diagram

$$\begin{array}{ccc} B^* & \xrightarrow{\ n\ } & K^* \\ \downarrow & & \downarrow \\ \hat{B}_y{}^* & \xrightarrow{\ n\ } & \hat{K}_y{}^* \end{array}$$

It will suffice to show that the image of $M_m(\mathbb{H})^* \xrightarrow{n} \mathbb{R}^*$ consists of the positive reals. But n is continuous and $M_m(\mathbb{H})^*$ is a connected topological group. Our assertion clearly follows from this. To see that $M_m(\mathbb{H})^*$ is connected let $A \in M_m(\mathbb{H})^*$. Reduce A to diagonal form $D = \text{diag}(d_1, \ldots, d_m)$ by elementary row and column operations. Then A and D are connected by a path in $M_m(\mathbb{H})^*$. For example, if we replace the i-th row r_i by $r_i + \alpha r_j$, $\alpha \in H$, we can instead replace r_i by $r_i + t\alpha r_j$, $0 \leq t \leq 1$. This gives a path between the initial and final values. Now $\mathbb{H}^* = \mathbb{R}^4 - \{0\}$ is connected. Let $d_i(t)$ be a path from d_i to 1 in $\mathbb{H}^*$. Then $D(t) = \text{diag}(d_i(t))$ gives a path from D to I.

The sufficiency of the condition will be proved in Chapter 9.

Recall now the following definition used in class field theory.

<u>Definition</u>. If $\mathfrak{m} = y_1 \ldots y_r$ is a formal product of real archimedian primes, the ray mod $\mathfrak{m}$ is defined to be $S^R_{\mathfrak{m}} = \{(a) | a > 0$ at $y_1, \ldots, y_r\}$, where $(a) = Ra$.

Since y ramifies in $M_s(A)$ if and only if it ramifies in A, Theorem 7.6 shows that for each s, the image Hs of $M_s(A)^*$ in I_R is precisely $S_{\mathfrak{m}}$ where $\mathfrak{m}$ is the (formal) product of all real archimedian primes ramified in A.

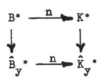

Theorem 7.8. Let R be a Dedekind ring with quotient field K and let Λ be a maximal R-order in a central simple K-algebra A. Assume K is a global field and let $\mathfrak{m}$ be the product of all real archimedean primes of K which ramify in A. Then we have an exact sequence

$$0 \to I_R/S^R_{\mathfrak{m}} \to K_0(\Lambda) \to K_0(A) \to 0$$

so $K_0(\Lambda) \approx \mathbb{Z} \oplus I_R/S^R_{\mathfrak{m}}$.

The class group of R is $Cl(R) = I_R/P_R$ where $P_R = \{(a) | a \in K^*\}$. If char K $\neq$ 0, then $I_R/S^R_{\mathfrak{m}} = Cl(R)$. If char K = 0, we have an exact sequence

$$0 \to P_R/S^R_{\mathfrak{m}} \to I_R/S^R_{\mathfrak{m}} \to Cl(R) \to 0 .$$

For real archimedean y, define $\hat{K}_y{}^* \to \mathbb{Z}/2\mathbb{Z}$ by $\hat{K}_y{}^* \cong \mathbb{R}^* \to \mathbb{R}^*/\mathbb{R}^{*+}$. Let $D = \prod \mathbb{Z}/2\mathbb{Z}$ over the $y | \mathfrak{m}$ and define $K^* \to D$ by $K^* \to \prod_{y|\mathfrak{m}} \hat{K}_y{}^* \to \prod_{y|\mathfrak{m}} \mathbb{Z}/2\mathbb{Z} = D$. This is onto by the approximation theorem. If X is the kernel of this, then $S^R_{\mathfrak{m}}$ is the image of X in P_R. Now $K^* \to P_R$ is onto and its kernel is clearly U(R), the group of units of R. Therefore we have $P_R/S^R_{\mathfrak{m}} = K^*/U(R)X = D/\text{im}(U(R) \to D)$. Thus we have an exact sequence

$$U(R) \to D \to I_R/S^R_{\mathfrak{m}} \to Cl(R) \to 0 .$$

Since D and Cl(R) are finite (by the Jordan-Zassenhaus theorem), so is $I_R/S^R_{\mathfrak{m}}$. (Of course this was already known by Theorem 3.8 .)

Chapter 8: K_1 and G_1

If R is any ring then $GL_n(R)$ is, by definition, the group of all invertible $n \times n$ matrices over R. We embed $GL_n(R)$ in $GL_{n+1}(R)$ by sending A to $\begin{pmatrix} A & \begin{matrix} 0 \\ \vdots \\ 0 \end{matrix} \\ 0 \dots 0 & 1 \end{pmatrix}$. Let $GL(R) = \varinjlim GL_n(R)$.

This is the group of infinite invertible matrices of the form

$$\begin{pmatrix} A & & & & 0 \\ & 1 & & & \\ & & 1 & & \\ & & & \ddots & \\ 0 & & & & \end{pmatrix}$$

where A is finite.

<u>Definition</u>. $K_1(R) = GL(R)/[GL(R), GL(R)]$.

We refer to SK Ch 13 for elementary properties of this and in particular for the fact that $[GL(R), GL(R)] = E(R)$, the subgroup of GL(R) generated by all elementary matrices of the form $1 + re_{ij}$, $r \in R$, $i \neq j$.

The main object of this chapter is to give a partial calculation of $K_1(\Lambda)$ for an order Λ. Before doing this, we will generalize some of the results of Chapter 7.

Let R be a Dedekind ring with quotient field K. Let Λ be an R-order in a semisimple separable K-algebra A. In Chapter 7 we determined the kernel of $K_0(\Lambda) \longrightarrow K_0(A)$ for the case of a maximal order. In the more general case where Λ is any order, this argument will not work because $K \otimes M \approx K \otimes N$ does not imply $M \sim N$. Let P be a finite non-empty set of primes such that Λ_y is a maximal order for $y \notin P$. Then $M_y \approx N_y$ for all $y \in P$ implies $M \sim N$. Therefore we consider $K_0(\Lambda) \longrightarrow \prod_{p \in P} K_0(\hat{\Lambda}_y)$ instead of $K_0(\Lambda) \rightarrow K_0(A)$.

More generally, let M be a finitely generated torsion free Λ-module. As in Chapter 6, we let D_M be the abelian group generated by all [N] where N is a direct summand of M^s for some s, with relations $[N_1 \oplus N_2] = [N_1] + [N_2]$. Thus $K_0(\Lambda) = D_\Lambda$. We will determine the kernel of $\mathcal{O}_M \longrightarrow \prod_{y \in P} D_{\hat{M}_y}$. If we let $\mathcal{O}_M$ be the full subcategory of all Λ-modules which are direct summands of M^s for some s, we can describe D_M as $K_0(D_M, \oplus)$.

As in Chapter 6, let J be the annihilator of KM in A and let $\bar{A} = A/J$, $\hat{\Lambda} = \Lambda / \Lambda \cap J$. Then M is a faithful $\hat{\Lambda}$-module. Since $\mathcal{O}_M$ and so D_M is the same whether we use Λ or $\hat{\Lambda}$, we can assume without loss of generality that M is a faithful Λ-module.

Let P be a finite non-empty set of primes of R such that Λ_y is a maximal order for $y \notin P$. Let $T_{\Lambda,P}$ be the category of finitely generated torsion Λ-modules with no y torsion for $y \in P$.

If $C \in T_{\Lambda,P}$ and if there is an exact sequence $0 \to N_2 \to N_1 \to C \to 0$ with $N_1, N_2 \in \mathcal{P}_M$, then $\gamma(C) = [N_1] - [N_2] \in D_M$ is well defined by Lemma 6.9.

<u>Lemma 8.1</u>. (1) If $C \in T_{\Lambda,P}$, there is an epimorphism $M^s \to C \to 0$ for some s.

(2) If $0 \to N_2 \to N_1 \to C \to 0$ with $C \in T_{\Lambda,P}$ and $N_1 \in \mathcal{P}_M$, then $N_2 \in \mathcal{P}_M$.

(3) If $N_1 \in \mathcal{P}_M$ and $N_2 \sim N_1$, then $N_2 \in \mathcal{P}_M$.

(4) If $0 \to C' \to C \to C'' \to 0$ in $T_{\Lambda,P}$, then $\gamma(C) = \gamma(C') + \gamma(C'')$ so γ gives a well defined homomorphism $\gamma\colon K_0(T_{\Lambda,P}) \to D_M$.

<u>Proof</u>. If C is simple, (1) is true by Lemma 6.7. We now use induction on the length of C. It will suffice to consider an exact sequence $0 \to C' \xrightarrow{i} C \xrightarrow{j} C'' \to 0$ and show that (1) for C' and C" implies (1) for C. More generally given epimorphisms $f'\colon N' \to C'$, $f''\colon N'' \to C''$, with $N', N'' \in \mathcal{P}_M$, we can lift f" to $f\colon N'' \to C$ such that

$$\begin{array}{ccccc} N'' & & & & \\ {\scriptstyle f}\downarrow & \searrow^{f''} & & & \\ C & \xrightarrow{j} & C'' & \longrightarrow & 0 \end{array}$$

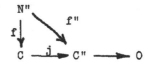

commutes. This argument was used in the proof of Lemma 6.9. It is now trivial to show that $(if', f): N' \oplus N'' \to C$ is an epimorphism. This completes the proof of (1). We also get a diagram

$$\begin{array}{ccccccccc} & & 0 & & 0 & & 0 & & \\ & & \downarrow & & \downarrow & & \downarrow & & \\ 0 & \to & N_1' & \to & N_1 & \to & N_1'' & \to & 0 \\ & & \downarrow & & \downarrow & & \downarrow & & \\ 0 & \to & N' & \to & N' \oplus N'' & \to & N'' & \to & 0 \\ & & \downarrow & & \downarrow & & \downarrow & & \\ 0 & \to & C' & \to & C & \to & C'' & \to & 0 \\ & & \downarrow & & \downarrow & & \downarrow & & \\ & & 0 & & 0 & & 0 & & \end{array}$$

If we have (2), the modules in the top row lie in $\mathcal{D}_M$. We claim the top row splits. By Lemma 6.14 we can check this locally. For $y \notin P$, Λ_y is maximal and so N''_{1y} is projective. If $y \in P$, $C_y = 0$ so the top row becomes isomorphic to the middle row locally at y. Since the top row splits, $[N_1] = [N_1'] + [N_1'']$ in D_M and we see immediately that (4) is satisfied.

If we localize the sequence in (2) at any $y \in P$, we get $N_{2y} \approx N_{1y}$ since $C_y = 0$. By Lemma 6.5, $N_2 \sim N_1$. Therefore (3) implies (2).

It remains to prove (3). We use the argument of Theorem 6.8. Since $N_1 \sim N_2$ we can find an exact sequence $0 \to N_1 \to N_2 \to C \to 0$ with $C \in T_{\Lambda, P}$ by Lemma 6.6 . By (1) we can find $0 \to P \to M^s \to C \to 0$. By Lemma 6.9, $N_2 \oplus P \approx N_1 \oplus M^s$. Since N_1 is a summand of M^t for some t, we see that N_2 is a summand of M^{s+t}.

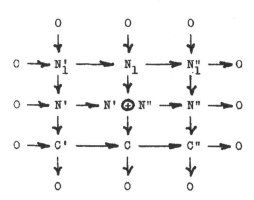

<u>Lemma 8.2</u>. $K_0(T_{\Lambda,P}) \xrightarrow{\gamma} D_M \longrightarrow \prod_{y \in P} D_{\hat{M}_y}$ is exact and ker γ is generated by all [C] with $C \in T_{\Lambda,P}$ such that there is an exact sequence $0 \to M^s \to M^s \to C \to 0$ for some s.

<u>Proof</u>. If $C \in T_{\Lambda,P}$, $0 \to N_2 \to N_1 \to C \to 0$ with $N_1, N_2 \in D_M$ then as above $N_1 \sim N_2$ so $[N_1] - [N_2] \mapsto 0$ in $D_{\hat{M}_y}$. Suppose conversely that $x = [N_1] - [N_2] \mapsto$ in $D_{\hat{M}_y}$ for $y \in P$. Then $N_1 \sim N_2$ by Lemma 6.5 and Theorem 5.20. As above we find $0 \to N_2 \to N_1 \to C \to 0$ with $C \in T_{\Lambda,P}$ using Lemma 6.6 . Therefore $x = \gamma(C)$.

If $0 \to M^s \to M^s \to C \to 0$ then clearly $\gamma(C) = 0$. Let $\gamma(x) = 0$ where $x = [C] - [D] \in K_0(T_{\Lambda,P})$. By Lemma 8.1 (1), find $0 \to N \to M^s \to D \to 0$. Since $D_y = 0$ for $y \in P$, there is an $r \neq 0$ in R with $rD = 0$, $y \nmid r$ for $y \in P$. (If $\mathfrak{a}$ is the annihilator of D, $\mathfrak{a} \subset \bigcup_P y$ would imply $\mathfrak{a} \subset y$ some $y \in P$ but $D_y = 0$.) Therefore, we have $0 \to D' \to M^s/rM^s \to D \to 0$ for some D'. Now $x = [C \oplus D'] - [D \oplus D'] = [C \oplus D'] - [M^s/rM^s]$. But $0 \to M^s \xrightarrow{r} M^s \to M^s/rM^s \to 0$ so $[M^s/rM^s]$ is one of our chosen generators. Also this shows $0 = \gamma(x) = \gamma[C \oplus D']$.

We now claim that $\gamma[C] = 0$ implies the existence of a sequence $0 \to M^s \to M^s \to C \to 0$. In fact, by Lemma 8.1 (1), find $0 \to N \to M^s \to C \to 0$. Now $\gamma[C] = [M^s] - [N] = 0$ so $M^s \oplus M^t \approx N \oplus M^t$ for some t and $0 \to N \oplus M^t \to M^s \oplus M^t \to C \to 0$. This proves the lemma.

Now let $S = R - \bigcup_{y \in P} y$. Then $C_y = 0$ for all $y \in P$ if and only if $C_S = 0$. If $0 \to M^s \xrightarrow{f} M^s \to C \to 0$ we see that f induces an automorphism of M^s_S. Conversely, if α is an automorphism of M^s_S, choose $t \in S$ and $f: M^s \to M^s$ so $\alpha = f/t$. Since $(\ker f)_S = 0$, ker f is a torsion module but M is torsion free. Thus f is a monomorphism. Let $C = \text{ckr } f$, $0 \to M^s \to M^s \to C \to 0$. Then $C_S = 0$ so $C \in T_{P,\Lambda}$. Also $M^s/tM^s \in T_{\Lambda,P}$. Define $d: \text{Aut}_{\Lambda_S}(M^s_S) \to K_0(T_{\Lambda,P})$ by $d(\alpha) = [C] - [M^s/tM^s]$. We claim that d is a well defined group homomorphism.

To prove this we use the following well known result.

Lemma 8.3. If $A \xrightarrow{f} B \xrightarrow{g} C$ then we have an exact sequence $0 \to \ker f \to \ker gf \to \ker g \to \text{ckr } f \to \text{ckr } gf \to \text{ckr } g \to 0$.

If we also have $\alpha = f'/t'$ then $t'f = tf'$. Apply Lemma 8.3 to these compositions and also to $t't = tt'$ to see that d is well defined. If $\beta = g/u$, Lemma 8.3 on fg and tu shows $d(\alpha\beta) = d(\alpha) + d(\beta)$.

Let $E = \text{End}_{\Lambda_S}(M_S)$. Then $\text{End}_{\Lambda_S}(M^s_S) = M_s(E)$ so $\text{Aut}_{\Lambda_S}(M^s_S) = GL_s(E)$ so we have a homomorphism $d: GL_s(E) \to K_0(T_{\Lambda,P})$. It is trivial to see that

$$\begin{array}{ccc} GL_s(E) & & \\ \downarrow & \searrow^{d} & \\ GL_{s+1}(E) & \xrightarrow{d} & K_0(T_{\Lambda,P}) \end{array}$$

commutes, so we have d: $GL(E) \to K_0(T_{\Lambda,P})$. Since K_0 is commutative, this induces $\partial: K_1(E) \to K_0(T_{\Lambda,P})$.

Proposition 8.4. Let Λ be an order over R in a semisimple separable K-algebra A. Let P be a finite non-empty set of primes of R such that Λ_y is a maximal order for $y \notin P$. Let M be a finitely generated Λ-module and $E = \mathrm{End}_{\Lambda_S}(M_S)$. Then we have an exact sequence

$$K_1(E) \to K_0(T_{\Lambda,P}) \to D_M \to \prod_{y \in P} D_{\hat{M}_y} .$$

This is immediate from the above results. An exact sequence of this type was considered by Heller and Reiner for ordinary K-theory. The generalization to D_M is due to Jacobinski. In Chapter 9 we will apply this result in a special case where E is semilocal (i.e., E/J is artinian where J is the Jacobson radical of E). In this case we can replace $K_1(E)$ by U(E) the group of units of E using

Proposition 8.5 (Bass). If E is a semilocal ring, then $U(E) = GL_1(E) \to K_1(E)$ is an epimorphism.

Proof. A more general result is proved in SK, Theorem 13.5. We will give the proof for the semilocal case again since it is very simple compared to the general case. By a very useful lemma of Bass SK Lemma 11.8, if $\mathfrak{a}$ is a right ideal of E and $Ex + \mathfrak{a} = E$, then there is an $a \in \mathfrak{a}$ such that $x + a = u$ is a unit of E.

Suppose $(a_{ij}) \in GL_n(E)$. Let $x = a_{11}$ and $\mathfrak{m} = a_{12}E + \dots + a_{1n}E$. We get a unit $u = a_{11} + a_{12}y_2 + \dots + a_{1n}y_n$. Thus by elementary row and column operations we reduce all other entries in the first row and column to 0. Repeating, we eventually reduce (a_{ij}) to a diagonal matrix. Since all $\begin{pmatrix} x & 0 \\ 0 & x^{-1} \end{pmatrix} \in E(E)$ by SK Lemma 13.3 we can reduce (a_{ij}) to $\begin{pmatrix} x_1 & & & \\ & 1 & & \\ & & \ddots & \\ & & & 1 \end{pmatrix} \in GL_1(E)$.

We now recall Bass' categorical definition of $K_1(R)$. We consider pairs (P, f) where P is a finitely generated projective R-module and f is an R-automorphism of P. A map $g: (P, f) \to (P', f')$ is a map $f: P \to P'$ such that the diagram

$$\begin{array}{ccc} P & \xrightarrow{g} & P' \\ {\scriptstyle f}\downarrow & & \downarrow{\scriptstyle f'} \\ P & \xrightarrow{g} & P' \end{array} \quad \text{commutes.}$$

Let $K_1(R)$ be the abelian group generated by all such $[(P, f)]$ with the relations

1) $[(P, f)] = [(P', f')] + [(P'', f'')]$ if there are maps $i: (P', f') \to (P, f)$ and $j: (P, f) \to (P'', f'')$ with $0 \to P' \xrightarrow{i} P \xrightarrow{j} P'' \to 0$ exact, i.e., $0 \to (P', f') \to (P, f) \to (P'', f'') \to 0$ is exact.

2) $[(P, fg)] = [(P, f)] + [(P, g)]$ where f and g are both automorphisms of P.

Theorem 8.6. $K_1(R)$ under the definition given is naturally isomorphic to the Whitehead group $GL(R)/[GL(R), GL(R)] = GL(R)/E(R)$ (the definition of $K_1(R)$ we have used previously).

This is one of the standard facts of K-theory. See SK Chapter 13.

As we shall see, the ability to switch back and forth between these definitions will be crucial in many proofs.

If R is a commutative ring then $K_0(R)$ is made into a ring by setting $[P][Q] = [P \otimes_R Q]$, and $K_1(R)$ is a $K_0(R)$ module by $[P][(Q, g)] = [(P \otimes_R Q, 1 \otimes g)]$. If R is commutative and Λ and Γ are R algebras then we can define $K_0(\Lambda) \times K_1(\Gamma) \to K_1(\Lambda \otimes_R \Gamma)$ by the same map, $[P][(Q, g)] = [(P \otimes_R Q, 1 \otimes g)]$.

If R is any ring and $R' = M_n(R)$ then $K_1(R) = K_1(R')$ since $M_k(R') = M_{kn}(R)$ and hence $GL_k(R') = GL_{kn}(R)$.

If A is a simple ring, then $A = M_n(D)$ for some division ring D and $K_1(A) = K_1(D)$. Let A be a central simple algebra over a field K. Then the reduced norm $n: M_r(A) \to K$ gives a homomorphism $GL_n(A) = M_n(A)^* \xrightarrow{n} K^*$. The diagram
$$\begin{array}{ccc} GL_n(A) & \searrow & \\ \downarrow & & K^* \\ GL_{n+1}(A) & \nearrow & \end{array}$$
commutes. In fact, if $L \supset K$ splits A then

$$\begin{array}{ccc} M_r(A)^* & \longrightarrow & (L \otimes_K M_r(A))^* = M_r(L \otimes_K A)^* \\ n \downarrow & & \swarrow n = \det \\ K^+ & \longrightarrow & L^* \end{array}$$

commutes by the definition of n, and

$$\begin{array}{ccc} GL_r(L \otimes_K A) & \xrightarrow{\det} & \\ \downarrow & & K^* \\ GL_{n+1}(L \otimes_K A) & \xrightarrow{\det} & \end{array}$$

clearly commutes. Thus n defines $n\colon K_1(A) \to K^*$. The image of this is known when K is an algebraic number field by the norm theorem. S. S. Wang has shown in the algebraic number field case the kernel is 0 (BK V, Theorem 9.7).

We now give a partial extension of this to arbitrary fields. A map $f\colon A \to B$ of abelian groups is called an isomorphism mod torsion if ker f and ckr f are torsion groups or equivalently if $1 \otimes f\colon Q \otimes A \xrightarrow{\approx} Q \otimes B$.

<u>Theorem 8.7</u>. Let K be any field and A a central simple K algebra. Then $n\colon K_1(A) \to K^*$ is an isomorphism mod torsion. In fact ker f and ckr f are annihilated by m where $m^2 = \dim_K A$.

<u>Proof</u>. The composite map $K^* \subset A^* = GL_1(A) \to K_1(A) \xrightarrow{n} K^*$ is just $x \leadsto x^m$ where $\dim_K A = m^2$. For let $L \supset K$ be a splitting field for A. Then

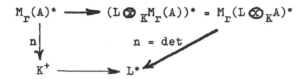

$$\begin{array}{ccc}
K^* & \longrightarrow & L^* \\
\downarrow & & \downarrow \\
A^* & \longrightarrow & (L \otimes_K A)^* = M_m(L)^* \\
\downarrow & & \downarrow \\
K_1(A) & \longrightarrow & K_1(L \otimes_K A) \\
n\downarrow & & \downarrow \text{det} \\
K^* & \longrightarrow & L^*
\end{array}$$

commutes.

But $x \in K^*$ goes to xI in $M_m(L)^*$ which goes to $x^m \in L^*$. But $K^+ \subset L^+$. Hence $x \rightsquigarrow x^m$ as desired. Thus the cokernel of $K_1(A) \to K^*$ is annihilated by m.

For the kernel let $L \supset K$ split A. Then $K_1(L \otimes_K A) = K_1(M_m(L)) = K_1(L) = L^*$. Hence if $x \in \mathrm{Ker}(K_1(A) \to K^*)$ then $x \in \mathrm{Ker}(K_1(A) \to K_1(L \otimes_K A))$. Thus the following lemma will complete the proof. We remark that if $A = M_s(D)$ with D a division ring, we may let L be a maximal subfield of D. Thus $[L: K] | M$.

<u>Lemma 8.8</u>. Let L be a finite extension of a field K of degree r. Then the kernel of $K_1(A) \to K_1(L \otimes_K A)$ has exponent dividing r.

<u>Proof</u>. Let $[(P, f)] \in K_1(L \otimes_K A)$ where P is a finitely generated projective $L \otimes_K A$ module and f is an automorphism. $L \otimes_K A$ is a free finitely generated A module since L is free and finitely generated over K.

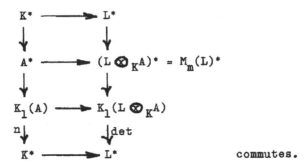

Hence, P is a finitely generated projective A module and f is an A automorphism, so we can define $[(P, f)] \in K_1(A)$. This yields a well defined map $K_1(L \otimes_K A) \to K_1(A)$. The composite $K_1(A) \to K_1(L \otimes_K A) \to K_1(A)$ is multiplication by $r = [L: K]$ since $L \otimes_K P \approx \coprod_1^r P$ as an A-module.

Thus if $x \in \ker[K_1(A) \to K_1(L \otimes_K A)]$ then $rx = 0$ in $K_1(A)$ as desired.

Corollary 8.9. If A is a semisimple algebra with center Z then the kernel and cokernel of $K_1(A) \xrightarrow{n} K_1(Z)$ are torsion groups of bounded exponent. ($K_1(A) = K_1(A_1) \times \ldots \times K_1(A_n)$ where $A = A_1 \times \ldots \times A_n$.)

We will now show that a similar result holds for orders.

Let R be a Dedekind ring with quotient field K, A a separable semisimple K algebra with center C, and Λ an order in A over R. Then we have maps $K_1(\Lambda) \to K_1(A) \xrightarrow{n} C^*$. Let R' be the integral closure of R in C. Then we claim the image of $K_1(\Lambda)$ in C* is contained in R'*. For everything is well behaved on products so we can assume that A is simple. Thus C is a field and R' a Dedekind ring. We can replace Λ by the bigger $R'\Lambda$, so we assume Λ is an R'-order. For any s, $M_s(\Lambda)$ is an R' order. Thus $n: M_s(\Lambda) \to R'$.

We can now state the main result which slightly generalizes a result of Bass.

Theorem 8.10. Let R be a Dedekind ring with quotient field K, A a separable semisimple K algebra, Λ an R order in A, C the center of

A and R' the integral closure of R in C. Then n: $K_1(L) \to R'^*$ is an isomorphism mod torsion if

1) $R/\mathfrak{a}$ is finite for every non zero ideal $\mathfrak{a}$, and

2) If L/K is a finite separable extension and R_1 is the integral closure of R in L, then the class group of R_1 is a torsion group.

Remark. This implies rk $K_1(\Lambda)$ = rk R^* for such orders.

We recall the definitions of the relative K_1. Let R be any ring and I a 2-sided ideal of R. Then GL(R, I) = ker[GL(R) $\to$ GL(R/I)]. We let E(R, I) be the smallest normal subgroup of E(R) containing all $I + qe_{ij}$ $i \neq j$ $q \in I$ and let $K_1(R, I) = GL(R, I)/E(R, I)$.

Proposition 8.11. Under the hypothesis of Theorem 8.10, if I is an ideal of Λ with KI = A, then $K_1(\Lambda, I) \to K_1(\Lambda)$ is an isomorphism mod torsion.

Proof. $GL(\Lambda/I) = \bigcup_{n=1}^{\infty} GL_n(\Lambda/I)$. But Λ/I is finite hence $GL_n(\Lambda/I)$ is finite for each n and $GL(\Lambda/I)$ is locally finite. Also $E(\Lambda) = [GL(\Lambda), GL(\Lambda)]$ and $E(\Lambda, I) = [GL(\Lambda), GL(\Lambda, I)]$. See SK, Theorem 13.2 and Theorem 15.1.

Hence it is enough to show the following lemma:

Lemma 8.12. Let G be a group and N a normal subgroup with G/N locally finite. Then $N/[G, N] \to G/[G, G]$ is an isomorphism mod torsion.

Proof. (Assuming results from cohomology of groups.)

$$0 \to N \to G \to G/N \to 0 \quad \text{is exact.}$$

Letting $H_n(G) = H_n(G, Z)$ we get an exact sequence

$H_2(G) \to H_2(G/N) \to H_0(G/N, H_1(N)) \to H_1(G) \to H_1(G/N) \to 0$. But $H_1(N) = N/[N, N]$, $H_0(G/N, H_1(N)) = N/[G, N]$, and $H_1(G) = G/[G, G]$ so $H_2(G) \to H_2(G/N) \to N/[G, N] \to G/[G, G] \to G/N[G, G] \to 0$. Hence it is enough to show that $H_1(G/N)$ and $H_2(G/N)$ are torsion. But G/N is locally finite. Hence $G/N = \varinjlim H_i$ with H_i finite. Therefore, $H_n(G/N) = \varinjlim H_n(H_i)$ and each $H_n(H_i)$ is finite. Thus, $H_1(G/N)$ and $H_2(G/N)$ are torsion. This completes the proof of proposition 8.11.

Keeping the above hypothesis. If Λ and Γ are R orders in A then so is $\Lambda \cap \Gamma$. If $K_1(\Lambda \cap \Gamma) \to K_1(\Lambda)$ and $K_1(\Lambda \cap \Gamma) \to K_1(\Gamma)$ are both isomorphisms mod torsion, the diagram

$$\begin{array}{ccccc} K_1(\Lambda) & \longleftarrow & K_1(\Lambda \cap \Gamma) & \longrightarrow & K_1(\Gamma) \\ & \searrow & \downarrow & \swarrow & \\ & & R^* & & \end{array}$$

shows that Theorem 8.10 holds for Λ if and only if it holds for Γ.

Lemma 8.13. Let G be a group and N a normal subgroup such that G/N is locally finite. Then $[G, G]/[G, N]$ is locally finite.

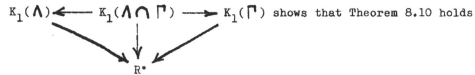

Proof. $[G, N] \subset N \cap [G, G] \subset [G, G]$. But $N \cap [G, G]/[G, N] \subset \ker[N/[G, N] \to G/[G, G]]$ and therefore is torsion and abelian by Lemma 8.12. Hence $N \cap [G, G]/[G, N]$ is locally finite. Now $[G, G]/N \cap [G, G]$ is the image of $[G, G]$ in G/N which is locally finite. Hence $[G, G]/N \cap [G, G]$ is also locally finite. Apply the following lemma.

Lemma 8.14. If B is a group and A a normal subgroup with A and B/A locally finite, then B is locally finite.

Proof. Let C be a finitely generated subgroup of B. Then the image $C/C \cap A$ of C in B/A is finitely generated and hence finite. Thus $C \cap A$ has finite index in C so $C \cap A$ is also finitely generated. Thus $C \cap A$ is finite since A is locally finite.

Proposition 8.15. With the hypothesis of Theorem 8.10, if $\Lambda \subset \Gamma$ are both R orders of A then $K_1(\Lambda) \to K_1(\Gamma)$ is an isomorphism mod torsion.

Proof. Pick $r \neq 0$ in R with $r\Gamma \subset \Lambda$ then $I = r\Gamma$ is a 2-sided ideal of both Λ and Γ. We claim that if J is a 2-sided ideal of both Λ and Γ then $GL(\Lambda, J) = GL(\Gamma, J)$. In fact, $GL_n(\Lambda, J)$ is the group of invertible matrices of the form $1 + Q$ where Q has entries in J. Thus it depends only on J and not on Λ.

We have a commutative diagram

$$\begin{array}{ccc} K_1(\Lambda, J) & \longrightarrow & K_1(\Gamma, J) \\ \downarrow & & \downarrow \\ K_1(\Gamma) & \longrightarrow & K_1(\Gamma) \end{array}$$

By Proposition 8.11 the columns are isomorphism mod torsion. Hence the bottom will be an isomorphism mod torsion if the top is.

As above, pick $r \neq 0$ in R with $r\Gamma \subset \Lambda$, let $I = r\Gamma$ and $J = I^2 = r^2\Gamma$. For any ideals A, B of Γ we have $E(\Gamma, AB) \subset [E(\Gamma, A), E(\Gamma, B)]$ since $1 + ab\, e_{ij} = [1 + ae_{ik}, 1 + be_{kj}]$.

Hence $E(\Gamma, J) \subset [E(\Gamma, I), E(\Gamma, I)] \subset [GL(\Gamma, I), GL(\Gamma, I)]$. But $GL(\Gamma, I) = GL(\Lambda, I)$. Hence $[GL(\Gamma, I), GL(\Gamma, I)] = [GL(\Lambda, I), GL(\Lambda, I)] \subset [GL(\Lambda), GL(\Lambda, I)] = E(\Lambda, I) \subset E(\Lambda)$.

The top row of * is

$$GL(\Lambda, J)/E(\Lambda, J) \twoheadrightarrow GL(\Gamma, J)/E(\Gamma, J) .$$

This map is onto since the GL's are the same and the kernel is $E(\Gamma, J)/E(\Lambda, J)$. But $E(\Gamma, J)/E(\Lambda, J) \subset E(\Lambda)/E(\Lambda, J)$ by the calculation above. Hence, it is enough to show that $E(\Lambda)/E(\Lambda, J)$ is locally finite.

Let $G = GL(\Lambda)$ and $N = GL(\Lambda, J)$. Then $G/N \subset GL(\Lambda/J)$ is locally finite since Λ/J is finite. Hence, by lemma 8.13, $[G, G]/[G, N]$ is locally finite. But $[G, G] = E(\Lambda)$ and $[G, N] = E(\Lambda, J)$. Hence $E(\Lambda)/E(\Lambda, J)$ is locally finite as desired.

This completes the proof of Proposition 8.15.

Now we return to the proof of Theorem 8.10. We pick a maximal order $\Gamma \supset \Lambda$. Then $K_1(\Lambda) \to K_1(\Gamma)$ is an isomorphism mod torsion by Proposition 8.15. Hence $K_1(\Lambda) \xrightarrow{n} R'^*$ will be an isomorphism mod torsion if $K_1(\Gamma) \xrightarrow{n} R'^*$ is.

Everything is now well behaved on products since if $A = A_1 \times \ldots \times A_r$, then $\Gamma = \Gamma_1 \times \ldots \times \Gamma_r$. Hence we can assume A is simple and Λ is a maximal order. Now C is a field and R' is a Dedekind ring finitely generated as an R-module. Since $\Lambda \subset R'\Lambda$ which is an order over R, we see that $\Lambda = R'\Lambda$ is an R' order in A. The hypotheses on R in Theorem 8.10 are inherited by R'. Therefore we can replace R' by R and assume that A is central simple over K (and so R' = R).

First we show the map is an epimorphism mod torsion. We examine the maps

$$\begin{array}{ccccc} R^* & \rightarrow & K_1(\Lambda) & \rightarrow & R^* \\ \downarrow & & \downarrow & & \downarrow \\ K^* & \rightarrow & K_1(A) & \rightarrow & K^* \end{array}$$

. The bottom composite sends x to x^m where $m^2 = \dim_K A$. Hence, so does the top. Thus $\mathrm{ckr}[K_1(\Lambda) \rightarrow R^*]$ has exponent dividing m.

To show the map is a monomorphism mod torsion we begin with

<u>Lemma 8.16</u>. Let R be a Dedekind ring, Λ an R algebra which is finitely generated as an R module, and R' a commutative ring which is finitely generated and projective as an R module. Then the map $K_1(\Lambda) \rightarrow K_1(R' \otimes_R \Lambda)$ is a monomorphism mod torsion.

<u>Proof</u>. If P is a finitely generated projective $R' \otimes_R \Lambda$ module, then P is also a finitely generated projective Λ module. The map $K_1(R' \otimes_R \Lambda) \rightarrow K_1(\Lambda)$ given by $[(P, f)] \rightsquigarrow [(P, f)]$ is clearly well defined. The composite $K_1(\Lambda) \rightarrow K_1(R' \otimes_R \Lambda) \rightarrow K_1(\Lambda)$ is

given by $[(P, f)] \rightsquigarrow [(R' \otimes_R P, 1 \otimes f)]$, which is nothing but multiplication by $[R'] \in K_0(R)$. We claim there is a non zero integer n and an $x \in K_0(R)$ such that $x[R'] = n[R]$. Then, if $y \in K_1(\Lambda)$ goes to 0 in $K_1(R' \otimes_R \Lambda)$, then $x[R']y = 0$ so $n[R]y = ny = 0$ in $K_1(\Lambda)$. Thus n annihilates the kernel of $K_1(\Lambda) \to K_1(R' \otimes_R \Lambda)$. Now $R' = F \oplus \underline{a}$ as an R module where F is a finitely generated free R-module and $\underline{a}$ is an ideal. Let $x = [F \oplus \underline{a}^{-1}]$. Then $x[R'] = [F \oplus \underline{a}^{-1}][F \oplus \underline{a}] = [(F \otimes F)] + [F \otimes \underline{a}] + [F \otimes \underline{a}^{-1}] + [\underline{a}^{-1} \otimes \underline{a}] = [(F \otimes F)] + [F][\underline{a} \oplus \underline{a}^{-1}] + [R] = [(F \otimes F)] + [F][R \oplus R] + [R] = n[R]$ for some non zero integer n as claimed. We use here the fact that $\mathfrak{a} \oplus \mathfrak{a}^{-1} \approx R \oplus R$, and $\mathfrak{a} \otimes \mathfrak{a}^{-1} \approx R$.

Let L be a finite separable field extension of K such that $L \otimes_K A = M_s(L)$ for some s. Let R' be the integral closure of R in L. Then R' satisfies the hypothesis of lemma 8.16.

The diagram

$$\begin{array}{ccc} K_1(\Lambda) & \longrightarrow & K_1(R' \otimes_R \Lambda) \\ \downarrow n & & \downarrow n \\ R^* & \longrightarrow & R'^* \end{array}$$

commutes.

But $K_1(\Lambda) \to K_1(R' \otimes_R \Lambda)$ is a monomorphism mod torsion. Thus, $K_1(\Lambda) \to R^*$ will be a monomorphism mod torsion if $K_1(R' \otimes_R \Lambda] \to R'^*$ is. Thus, we can assume that $A = M_n(K)$ for some n. By the remark just before Lemma 8.13 we can also assume $\Lambda = M_n(R)$. Now $K_1(M_n(R))$ is naturally isomorphic to $K_1(R)$ as we have already observed. This isomorphism, given by $M_k(M_n(R)) \approx M_{nk}(R)$ clearly preserves $n = \det: K_1 \to R^*$. Thus it is enough to prove the theorem for $\det: K_1(R) \to R^*$.

By SK Theorem 13.5 the map $GL_2(R)$ to $K_1(R)$ is onto. Let $x \in K_1(R)$ go to 1 in R^*.

We claim we can pick $\begin{pmatrix} a & b \\ c & d \end{pmatrix} \in GL_2(R)$ representing x such that a separable field extension L of K contains a root of the characteristic polynomial of $\begin{pmatrix} a & b \\ c & d \end{pmatrix}$. If $\mathrm{char}(K) \neq 2$, this is automatic. If $\mathrm{char}(K) = 2$, then the characteristic polynomial is separable unless $a + d = 0$. If $a + d = 0$, then we modify $\begin{pmatrix} a & b \\ c & d \end{pmatrix}$ by the elementary matrix $\begin{pmatrix} 1 & 1 \\ 0 & 1 \end{pmatrix}$ which does not change the image in $K_1(R)$. $\begin{pmatrix} a & b \\ c & d \end{pmatrix}\begin{pmatrix} 1 & 1 \\ 0 & 1 \end{pmatrix} = \begin{pmatrix} a & a+b \\ c & d+c \end{pmatrix}$. The latter has separable characteristic polynomial unless $c = 0$. If $c = 0$ and $b \neq 0$, then $\begin{pmatrix} 1 & 0 \\ 1 & 1 \end{pmatrix}\begin{pmatrix} a & b \\ 0 & d \end{pmatrix} = \begin{pmatrix} a & b \\ 0 & b+d \end{pmatrix}$ has a separable characteristic polynomial. If $b = c = 0$, then $a^{-1} = d$ since $\det\begin{pmatrix} a & 0 \\ 0 & d \end{pmatrix} = \det x = 1$. But $\begin{pmatrix} a^{-1} & 0 \\ 0 & a \end{pmatrix} \in E(R)$ by SK, Lemma 13.3. Thus $\begin{pmatrix} a & 0 \\ 0 & a^{-1} \end{pmatrix}$ is equivalent to the identity which has its characteristic roots in K. Of course $x = 1$ in this case. Let R' be the integral closure of R in L. Then lemma 8.16 applies to the map $K_1(R) \to K_1(R')$. Since

the diagram
$$\begin{array}{ccc} K_1(R) & \longrightarrow & K_1(R') \\ {\scriptstyle\det}\downarrow & & \downarrow{\scriptstyle\det} \\ R^* & \longrightarrow & R'^* \end{array}$$
commutes, we can assume K contains a root of the characteristic polynomial of $\begin{pmatrix} a & b \\ c & d \end{pmatrix}$.

Thus $x_1 = \begin{pmatrix} a & b \\ c & d \end{pmatrix}$ has an eigenvalue in K. We break K^2 up as $K_1 \oplus K_2$ where K_1 is generated by u with $x_1 u = \lambda u$, $\lambda \in K$ and hence is stable under x_1 and x_1^{-1}. Then $P = R^2 \cap K_1$ is stable under x_1 and x_1^{-1}. Passing to the categorical definition of K_1 we have a commutative diagram of exact sequences

$$\begin{array}{ccccccccc} 0 & \longrightarrow & P & \longrightarrow & R^2 & \longrightarrow & Q & \longrightarrow & 0 \\ & & {\scriptstyle f}\downarrow & & \downarrow{\scriptstyle x_1} & & \downarrow{\scriptstyle g} & & \\ 0 & \longrightarrow & P & \longrightarrow & R^2 & \longrightarrow & Q & \longrightarrow & 0 \end{array}$$

where f and g are induced from x_1 and are isomorphisms. Then $x = [(R^2, x_1)] = [(P, f)] + [(Q, g)]$. P and Q are rank 1 torsion free finitely generated. Hence P is isomorphic to $\underline{a}$ and Q to $\underline{b}$ where $\underline{a}$ and $\underline{b}$ are ideals and f and g must be multiplication by units r and s of R.

x goes to 1 in K^*. Therefore, rs = 1. Hence, $[(\underline{b}, \hat{s})] = [(\underline{b}, \widehat{r^{-1}})] = -[(\underline{b}, \hat{r})]$. Thus $x = ([\underline{a}] - [\underline{b}])([R, \hat{r}])$ where $K_1(R)$ is considered as a $K_0(R)$ module. But $[\underline{a}]-[\underline{b}] \in C_0(R)$, the class group of R, which is a torsion group by hypothesis.

Hence, $x \in \text{image}(C_0(R) \to K_1(R))$ given by $z \rightsquigarrow z\,[(R, \hat{r})]$. But this image is a torsion group. Hence has finite order. This completes the proof of Theorem 8.10.

Corollary 8.17. (Bass). Let Q be the rational numbers, Z the integers, A a semisimple Q algebra, and Λ a Z order in A. Let q be the number of simple components of $A = A_1 \times \dots \times A_q$, and r the number of simple components of $\mathbb{R} \otimes_Q A$ where $\mathbb{R}$ is the real numbers. Then rank $K_1(\Lambda) = r - q$.

Proof. Let C be the center of A. Then $C = K_1 \times \dots \times K_q$. Let R_i be the ring of integers of K_i and $R = R_1 \times \dots \times R_q$. Then by Theorem 8.10 rank $K_1(\Lambda) = \sum_{i=1}^{q} \text{rank } R_i^*$. By Dirchlet's unit theorem (W, OM), the rank R'^*_i is one less than the number of archimidean primes of K_i. Now $\mathbb{R} \otimes_Q K_i = L_1 \times \dots \times L_{n_i}$ with each $L_v \approx \mathbb{R}$ or $\mathbb{C}$ and the number of archimidean primes of K_i is n_1. Thus $\sum \text{rk} R_i^* = \sum (n_i - 1) = \sum n_i - q$. But the number of simple components of $\mathbb{R} \otimes_Q A_i$ is n_i.

Corollary 8.18. Let G be a finite group. Then rank $K_1(ZG)$ = the number of irreducible real representations of G — the number of irreducible rational representations of G.

Remark. If Λ is an order over $\mathbb{Z}$, a theorem of Siegel shows that $GL_n(\Lambda)$ is finitely generated for all n. By SK, Theorem 13.5, $GL_2(\Lambda) \to K_1(\Lambda)$ is finitely generated. Thus $K_1(\Lambda)$ is finitely generated. (See BK X, Theorem 3.2.)

We will now obtain a result similar to Theorem 8.10 for $G_1(\Lambda)$.

Definition. Let Λ be a left noetherian ring. Then $G_1(\Lambda)$ is the abelian group defined by generators $[(M, f)]$ where f is an automorphism of the finitely generated left Λ module M with the relations
(1) for every exact sequence $0 \to (M', f') \to (M, f) \to (M'', f'') \to 0$ we have $[(M, f)] = [(M', f')] + [(M'', f'')]$ and
(2) $[(M, fg)] = [(M, f)] + [(M, g)]$.

The definition is just like that of K_1 except for the choice of the modules M. Thus there is a natural map $\kappa: K_1(\Lambda) \to G_1(\Lambda)$ by $\kappa[(P, f)] = [(P, f)]$.

Theorem 8.19. Let Λ be a left noetherian ring and I a nilpotent 2 sided ideal. Then we have an isomorphism $G_1(\Lambda/I) \xrightarrow{\approx} G_1(\Lambda)$ given by $[(M, f)] \rightsquigarrow [(M, f)]$. (This makes sense since each Λ/I module M can be regarded as a Λ module.)

We claim the map $G_1(\Lambda) \to G_1(\Lambda/I)$ given by $[(M, f)] \rightsquigarrow \sum_i [(M_i/M_{i+1}, f)]$ is the inverse of $G_1(\Lambda/I) \to G_1(\Lambda)$.

By the Jordan-Holder-Zassenhaus theorem any two such filtrations will have a common refinement. Since the Jordan-Holder-Zassenhaus construction uses just sums and intersections, the refinement will still be stable under f and f^{-1}. Hence 1) and 2) will be preserved. Therefore, to prove the map is well defined it is enough to show that adding one extra term to the filtration does not change the image. Say $M_i \supset N \supset M_{i+1}$. Then $0 \to N/M_{i+1} \to M_i/M_{i+1} \to M_i/N \to 0$ is exact and $[(M_i/M_{i+1}, f)] = [(N/M_{i+1}, f)] + [(M_i/N, f)]$. Thus the map is well defined.

If $0 \to (M', f') \to (M, f) \to (M'', f'') \to 0$ is exact, take a filtration for M' and M" and paste together to get one for M. Then that filtration shows the relations (1) are preserved. For (2), use $M \supset IM \supset \ldots \supset I^n M = 0$, for $[(M, fg)]$, $[(M, f)]$ and $[(M, g)]$. This shows relations (2) are preserved.

<u>Theorem 8.20</u> (Lam). Let R be a Dedekind ring with quotient field K and A be a separable semisimple K algebra such that

1) R/p is finite for every non-zero prime ideal $p \subset R$, and

2) If L is a finite separable field extension of K and R' is the integral closure of R in L, then the class group of R' is a torsion group. Then 1) If Λ is an R order in A, the map $\chi: K_1(\Lambda) \to G_1(\Lambda)$ is an isomorphism mod torsion,

2) If Γ is an R order in A containing Λ, then the map $G_1(\Gamma) \to G_1(\Lambda)$ is an isomorphism mod torsion,

3) If C is the center of A and R' is the integral closure of R in C, then the composite $G_1(\Lambda) \to G_1(A) = K_1(A) \xrightarrow{n} R'^*$ is an isomorphism mod torsion, and

4) If $K_1(\Lambda)$ is finitely generated for all maximal R-orders Λ of A, then $G_1(\Lambda)$ is finitely generated for <u>all</u> R orders Λ in A.

<u>Proof</u>. If $S = R - \{0\}$, then localizing at S is an exact functor and gives a map $G_1(\Lambda) \to G_1(\Lambda_S) = G_1(A)$. Since A is semisimple, every module is projective and so $G_1(A) = K_1(A)$. We have a commutative diagram

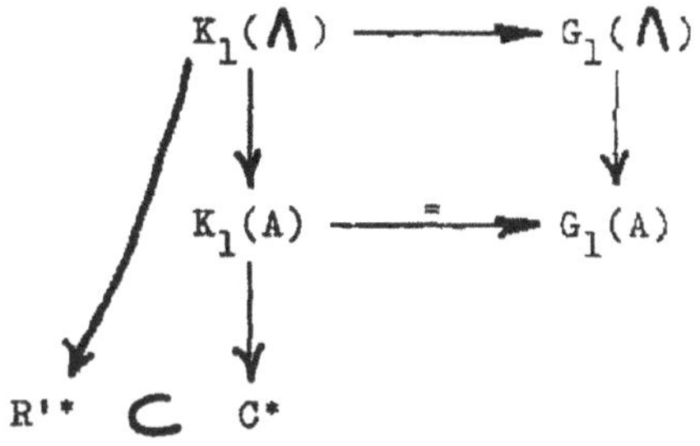

where C is the center of A and R' is the integral closure of R in C. By Theorem 8.10 the map $K_1(\Lambda) \to R'^*$ is an isomorphism mod torsion. Hence $\ker(K_1(\Lambda) \to G_1(\Lambda))$ is torsion.

Let $\Lambda \subset \Gamma$ with Γ a maximal order. Choose $r \neq 0$ in R with $r\Gamma = I \subset \Lambda$ then we have maps $G_1(\Lambda/I) \to G_1(\Lambda)$ and

$G_1(\Gamma) \to G_1(\Lambda)$. We claim $G_1(\Gamma) \oplus G_1(\Lambda/I) \to G_1(\Lambda)$ is onto. By lemma 8.22 below, $G_1(\Lambda)$ is generated by the [(M, f)] where M is torsion free as an R module. For such an (M, f), $M \subset KM = K \otimes_R M$ and f induces an automorphism of KM and hence also of $\Gamma M \subset KM$ since ΓM is stable under f and f^{-1}. We examine the short exact sequence of Λ modules $0 \to M \to \Gamma M \to \Gamma M/M \to 0$. Now $I\Gamma M = r\Gamma\Gamma M = r\Gamma M \subset \Lambda M = M$, so $\Gamma M/M$ is annihilated by I. f induces automorphism f' and f" on ΓM and $\Gamma M/M$ respectively and $[(M, f)] = [(\Gamma M, f')] - [(\Gamma M/M, f")]$ which is in the image of $G_1(G) \oplus G_1(L/I)$ as claimed. Now $\hat{\Lambda} = \Lambda/I$ is finite since $r\Lambda \subset I \subset \Lambda$. Thus $\hat{\Lambda}$ has a radical $\underline{a}$ which is nilpotent. Let $\bar{\bar{\Lambda}} = \Lambda/\underline{a}$. Then $G_1(\bar{\Lambda}) = G_1(\bar{\bar{\Lambda}})$ by Theorem 8.19. Since Λ is semisimple, $G_1(\bar{\bar{\Lambda}}) = K_1(\bar{\bar{\Lambda}})$ and furthermore, $\bar{\bar{\Lambda}}^* \to K_1(\bar{\bar{\Lambda}})$ is onto. Hence $K_1(\bar{\bar{\Lambda}})$ is finite since $\bar{\bar{\Lambda}}$ is. Thus, if Γ is a maximal order the map $G_1(\Gamma) \to G_1(\Lambda)$ has a finite cokernel since $G_1(\Lambda/I)$ is finite. Let Λ' be any order between Γ and Λ. Then we have $G_1(\Gamma) \to G_1(\Lambda') \to G_1(\Lambda)$ so the map $G_1(\Lambda') \to G_1(\Lambda)$ has a finite cokernel since $G_1(\Gamma) \to G_1(\Lambda)$ does. Since maximal orders are hereditary, SK, Th. 16.11 shows that $K_1(\Gamma) = G_1(\Gamma)$. Thus $G_1(\Lambda)$ is finitely generated if $K_1(\Gamma)$ is. This establishes 4).

Now if (3) is true, and $\Lambda \subset \Gamma$, we have a commutative diagram

$$\begin{array}{ccc} G_1(\Gamma) & \searrow & \\ \downarrow & & R'^* \\ G_1(\Lambda) & \nearrow & \end{array}$$

and the maps $G_1(\Gamma) \to R'^*$ and $G_1(\Lambda) \to R'^*$ are isomorphism mod torsion. Hence so is $G_1(\Gamma) \to G_1(\Lambda)$. Hence 3) implies 2).

Also, the map $K_1(\Lambda) \to R'^*$ is an isomorphism mod torsion by Theorem 8.10. If (3) holds the map $G_1(\Lambda) \to R'^*$ is an isomorphism mod torsion. Hence $K_1(\Lambda) \to G_1(\Lambda)$ is an isomorphism mod torsion. Thus (3) implies (1).

Therefore we only have to prove (3). We claim it is enough to verify (3) for maximal orders. If $\Lambda \subset \Gamma$, we have the commutative diagram

$$\begin{array}{ccc} G_1(\Gamma) & \searrow & \\ \downarrow & & R'^* \\ G_1(\Lambda) & \nearrow & \end{array}$$

and $G_1(\Gamma) \to G_1(\Lambda)$ is an epimorphism mod torsion. If (3) is true for Γ, then $G_1(\Gamma) \to G_1(\Lambda)$ is a monomorphism mod torsion since the diagram commutes. Hence $G_1(\Gamma) \to G_1(\Lambda)$ and $G_1(\Gamma) \to R'^*$ are both isomorphisms mod torsion and thus, $G_1(\Lambda) \to R'^*$ is also.

Hence we can assume $\Lambda = \Gamma$ is a maximal order. Then $K_1(\Gamma) = G_1(\Gamma)$ since maximal orders are hereditary. But

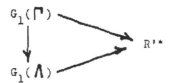

$K_1(\Gamma) \to R'^*$ is an isomorphism mod torsion by Theorem 8.10.

Proposition 8.21. Let R be a Dedekind ring with quotient field K, A be a semisimple K algebra, and Λ an R order in A. Let $A = A_1 \times \dots \times A_n$ be the decomposition of A into simple components and let Λ_i be the image of Λ in A_i so that $\Lambda \subset \Gamma = \Lambda_1 \times \dots \times \Lambda_n$. Then the map $G_1(\Gamma) \to G_1(\Lambda)$ is onto.

Proof. By Lemma 8.22, $G_1(\Lambda)$ is generated by $[(M, f)]$ with M a torsion free finitely generated Λ module.

Now $K \otimes_R M$ is an A module. Hence $K \otimes_R M = N_1 \oplus \dots \oplus N_n$ where each N_i is an A_i module and the N_i are fully invariant. We filter $K \otimes_R M$ by $K \otimes_R M = V_1 \supset V_2 \supset \dots \supset V_r \supset 0$ where $V_i = N_i \oplus N_{i+1} \oplus \dots \oplus N_r$. Then V_i/V_{i+1} is isomorphic to N_i. Let $M_i = M \cap V_i$ and $f_i = f/_{M_i}$. This is an automorphism of M_i since M_i is clearly stable under f and f^{-1}. Now $M = M_1 \supset M_2 \supset \dots \supset M_r \supset 0$ and $[(M_1 f)] = \sum [(M_i/M_{i+1}, \bar{f}_i)]$ where $\bar{f}_i$ is the map induced on M_i/M_{i+1} by f. Hence it is enough to show that all $[(M_i/M_{i+1}, \bar{f}_i)]$ are in the image of $G_1(\Gamma)$. Now A_j annihilates V_i/V_{i+1} unless $i = j$. Hence V_i/V_{i+1} is an A_i module and Λ acts on M_i/M_{i+1} through the projection $\Lambda \to \Lambda_i$. Thus Γ acts on M_i/M_{i+1} through the projection $\Gamma \to \Lambda_i$. Hence $[(M_i/M_{i+1}, \bar{f}_i)]$ is in $G_1(\Gamma)$ and the map $G_1(\Gamma) \to G_1(\Lambda)$ is onto as desired.

The following result follows from SK, Chapter 16, but we include a proof for the reader's convenience.

Lemma 8.22. Let R be a Dedekind ring and Λ an R-algebra finitely generated and torsion free as an R-module. Then $G_1(\Lambda)$ is generated by the [(M, f)] where M is finitely generated and torsion free as an R-module.

Proof. Let [(M, f)] be any generator of $G_1(\Lambda)$. Let $\eta: Q \to M$ be an epimorphism with Q projective over Λ. Let $P = Q \oplus Q$ and consider $P = Q \oplus Q \xrightarrow{pr_1} Q \to M$. This factors as $P = Q \oplus Q \xrightarrow{\eta \oplus \eta} M \oplus M \xrightarrow{pr_1} M$. Lift f to the automorphism $\begin{pmatrix} f & 0 \\ 0 & f^{-1} \end{pmatrix}$ of $M \oplus M$. This is a product of elementary automorphisms $\begin{pmatrix} 1 & a \\ 0 & 1 \end{pmatrix}$ and $\begin{pmatrix} 1 & 0 \\ b & 1 \end{pmatrix}$. These lift to automorphisms of $Q \oplus Q$ since Q is projective. Thus we can lift f to an automorphism g of P. The sequence $0 \to (N, b) \to (P, g) \to (M, f) \to 0$ shows that $[(M, f)] = [(P, g)] - [(N, b)]$ and N, P are torsion free.

Remark. With a bit more work we can show $G_1(\Lambda)$ is K_1 of the category of finitely generated torsion free Λ-modules. See SK, Corollary 16.21.

Chapter 9: Cancellation Theorems

In this chapter we will be concerned with the following problem. If $X \oplus M \cong X \oplus N$, is $M \cong N$? As always, we will consider finitely generated torsion free Λ-modules where Λ is an order over a Dedekind ring R in a semisimple separable K-algebra A, K being the quotient field of R. We will give two partial answers due to Bass and Jacobinski. Jacobinski's result depends on a deep theorem of Eichler and most of this chapter will be concerned with the proof of Eichler's theorem.

In SK, Chapter 12 we gave Bass' original proof of his theorem. Here we will give a different proof which shows more clearly the relation between the results of Bass and Jacobinski. The approach here is much more specialized since it depends on the fact that Λ is an order. We do not require the modules to be projective here. However, Dress [DR] has shown that the original proofs of Serre and Bass can be modified to avoid this assumption. We will illustrate our approach by first proving Dress' extension of Serre's theorem for the case of orders.

The hypotheses on R, K, A, Λ stated above will be in force throughout this chapter and will not be repeated. Let P be a finite non empty set of primes such that Λ_y is maximal for $y \notin P$.

<u>Theorem 9.1</u>. Let M and N be torsion free finitely generated Λ-modules. If $\hat{N}_y \oplus \hat{N}_y$ is a direct summand of $\hat{M}_y$ for all $y \in P$, then N is a direct summand of M.

<u>Proof</u>. By Theorem 6.12, $M = N_1 \oplus M'$ where $N_1 \sim N_1$. Since the Krull-Schmidt theorem holds over $\hat{\Lambda}_y$, N_y is a direct summand of $\hat{M}'_y$ for $y \in P$. By Theorem 6.12 again, $M' = N_2 \oplus P$ where $N_2 \sim N$. By

Roiter's lemma we can embed N_1 and N_2 in N so that N/N_1, N/N_2 have relatively prime orders. Thus $N_1 + N_2 = N$ so we have an exact sequence

$$0 \to Q \to N_1 \oplus N_2 \to N \to 0 .$$

For any y either $(N/N_1)_y = 0$ or $(N/N_2)_y = 0$. Therefore $N_{1y} = N_y$ or $N_{2y} = N_y$ so the sequence splits locally. By Lemma 6.14 it splits and so $N_1 \oplus N_2 \approx N \oplus Q$. But $M = N_1 \oplus N_2 \oplus P = N \oplus Q \oplus P$.

We now turn to the cancellation problem. We have $X \oplus M \approx X \oplus N$ where X, M, N are finitely generated torsion free Λ-modules. In both theorems to be proved we must assume X locally a direct summand of M^n for some n. Note that this is true in Bass' original theorem in which X is projective and M has a free summand. By Theorem 6.12, this implies $X \sim X'$ where X' is a direct summand of M^n. By Lemma 8.1 (3), $X \in D_M$ so X itself is a direct summand of some $M^s = X \oplus Y$. We conclude that $M^s \oplus M \approx M^s \oplus N$ so $N \in D_M$. Since $X \oplus M \approx X \oplus N$, we have $[M] - [N] = 0$ in D_N.

Let P be a finite non empty set of primes such that Λ_y is a maximal order for $y \notin P$. For Jacobinski's theorem we will also have to assume that certain other primes lie in P. By Roiter's lemma we can embed N in M such that $C = M/N$ has $C_y = 0$ for $y \in P$. (Clearly $N \sim M$ by Lemma 6.4.) Thus $C \in T_{\Lambda,P}$ using the notation of Proposition 8.5. Now $[C] \mapsto [M] - [N] = 0$ under the map $K_0(T_{\Lambda,P}) \to D_M$. By Proposition 8.5 there is an $\alpha \in K_1(\mathrm{End}_{\Lambda_S}(M_S))$ such that $\alpha \mapsto [C]$.

Now $S = R - \bigcup_{y \in P} y$ so R_S is semilocal having as maximal ideals the y_S for $y \in P$. By the next lemma, $E + \mathrm{End}_{\Lambda_S}(M_S)$ is semilocal since E is finitely generated as an R_S module.

Lemma 9.2. If R is a semilocal commutative ring and E is an R-algebra finitely generated as an R-module, then E is semilocal.

Proof. By definition, this means that E/J is artinian where J is the Jacobson radical of E. Let $\mathfrak{M}_1, \ldots, \mathfrak{M}_n$ be the maximal ideals of R and $\mathfrak{A} = \mathfrak{M}_1 \cap \ldots \cap \mathfrak{M}_n$ its Jacobson radical. Then $E/\mathfrak{A}E = \prod E/\mathfrak{M}_i E$ by the Chinese Remainder Theorem and $E/\mathfrak{M}_i E$ is a finite dimensional algebra over a field $R/\mathfrak{M}_i$. Therefore it will suffice to show that $\mathfrak{A} E \subset J$. For this we need only show that $\mathfrak{A}$ annihilates all simple Λ-modules. If S is a simple Λ-module and $\mathfrak{A} S \neq 0$ then $\mathfrak{A} S = S$ so $\mathfrak{M}_i S = S$ for all i. Localize at $\mathfrak{M}_i$ and use Nakayama's lemma to conclude $S_{\mathfrak{M}_i} = 0$. Thus $S = 0$, a contradiction.

Since E is semilocal, the map $U(E) = GL_1(E) \to K_1(E)$ is onto by Proposition 8.5. Therefore we can choose $\alpha \in U(E) = \mathrm{Aut}_{\Lambda_S}(M_S)$. The map $K_1(E) \to K_0(T_{\Lambda, P})$ was constructed as follows. If $\alpha \in GL_\Lambda(E)$, write $\alpha = f/s$ where $s \in S$ and $f \in M_n(\mathrm{End}_\Lambda(M))$. We get $0 \to M^n \xrightarrow{f} M^n \to D \to 0$ and $0 \to M^n \xrightarrow{s} M^n \to E \to 0$ and the image of α in $K_0(T_{\Lambda\, P})$ is [D] - [E]. In our case, n = 1 and also [D] - [E] = [C] so [D] = [C $\oplus$ E]. Therefore D and C $\oplus$ E have

composition series with isomorphic factors. Now we have
$0 \to N \xrightarrow{i} M \to C \to 0$ and $0 \to M \xrightarrow{s} M \to E \to 0$. Therefore if we define H by $0 \to N \xrightarrow{s_1} M \to H \to 0$ we have $0 \to C \to H \to E \to 0$ so H and D have isomorphic composition factors and $0 \to M \xrightarrow{f} M \to D \to 0$, $0 \to N \to M \to H \to 0$.

We can now reduce to the case where H = D and is simple by using the following lemma.

Lemma 9.3. If D is a torsion Λ module with $D_y = 0$ for $y \in P$, we can find a composition series for D in which the composition factors appear in any pre-assigned order.

Proof. If $D = D_1 \oplus D_2$ and the lemma holds for D_1 and D_2, then it clearly holds for D. Since D is the sum of its y-primary components, we can assume D is y-primary. Therefore $D = D_y$ and we can regard D as a module over Λ_y. Since Λ_y is a maximal order, we are reduced to the case where Λ is maximal. In this case $\Lambda = \Lambda_1 \times \dots \times \Lambda_s$ where $K\Lambda_i = A_i$, $A = A_1 \times \dots \times A_s$ with A_i simple. Now $D = D_1 \times \dots \times D_s$ where D_i is a Λ_i module. It will suffice to consider each D_i. Thus we can replace Λ by Λ_i. In other words we need only do the case where Λ is a maximal order in a simple algebra A. If R' is the integral closure of R in the center of A then $R'\Lambda$ is an order containing Λ (Note that all our algebras are assumed separable so R' is a finitely generated R-algebra). Thus $\Lambda = R'\Lambda$ so Λ is an order over R'. Since R does not occur in the statement of the lemma we can replace R by R' so we can assume A is central simple over K. Repeating the first part of the argument we reduce to the case where R is a DVR. Let y be its maximal ideal. By Theorem 5.16, there is a unique prime ideal (= maximal 2 sided ideal) P of Λ with $y\Lambda \subset P$ and $P^e \subset y\Lambda$ for some e. If S is a simple Λ-module then yS = 0 (If yS = S then S = 0 by Nakayama's lemma). Therefore PS = 0 so S is a Λ/P-module. But Λ/P is simple since P is a maximal 2-sided ideal and so has only one simple module up to isomorphism. Therefore the same is true for Λ and any composition series will do.

We now return to our sequences $0 \to M \to M \xrightarrow{\varphi} D \to 0$ and $0 \to N \to M \xrightarrow{\psi} H \to 0$. Since D and H have the same composition factors, Lemma 9.3 gives us composition series $D = D_0 > D_1 > \dots > D_h = 0$ and $H = H_0 > H_1 > \dots > H_h = 0$ with $D_i/D_{i+1} \approx H_i/H_{i+1}$. Let $M_i = \varphi^{-1}(D_i)$, $N_i = \psi^{-1}(H_i)$, and set $S_i = D_i/D_{i+1}$. Then we have

$$0 \to M_{i+1} \to M_i \to S_i \to 0$$

$$0 \to N_{i+1} \to N_i \to S_i \to 0$$

and $M_0 = M$, $N_0 = M$, $M_h \approx M$, $N_h \approx N$.

This leads at once to the following result.

Proposition 9.4. Let M be a finitely generated torsion free Λ-module with the following property: There is a finite set P of primes of R such that if $M' \sim M$ and S is a simple Λ-module with $S_y = 0$ for all $y \in P$, then any two epimorphisms $M' \twoheadrightarrow S$ have isomorphic kernels. Let X be a finitely generated Λ-module which is a local direct summand of M^n for some n. Then $X \oplus M \approx X \oplus N$ implies $M \approx N$.

Proof. We can enlarge P without spoiling the hypothesis and so assume that Λ_y is maximal for $y \notin P$. The construction above now applies. Since $(S_i)_y = 0$ for $y \in P$, we have $M_i \sim M_{i+1}$ and $N_i \sim N_{i+1}$. Therefore $M_i \sim M$, $N_i \sim M$ for all i. If $M_i \approx N_i$, the hypothesis implies $M_{i+1} \approx N_{i+1}$. But $M_0 = M = N_0$. Therefore $M \approx M_h \approx N_h \approx N$.

We must now prove the hypothesis of Proposition 9.4 holds under certain conditions. We do this by showing that if $f, g: M' \twoheadrightarrow S$ are epimorphisms then $g = f\theta$ for some automorphism θ of M'. This clearly implies $\ker f \approx \ker g$.

Proposition 9.5. Let P be a finite set of primes such that Λ_y is maximal for $y \notin P$. Let M be a finitely generated torsion free Λ-module such that $M = M_1 \oplus M_2$ where M_1 and M_2 are faithful Λ-modules. If S is a simple Λ-module and $f, g: M \twoheadrightarrow S$ are epimorphisms, then there is an automorphism θ of M with $g = f\theta$.

Proof. The relation "$g = f\theta$ for some automorphism θ" is clearly an equivalence relation. We will show any epimorphism $f: M \twoheadrightarrow S$ is equivalent to one particular $f_0: M \twoheadrightarrow S$. By Lemma 6.7 there are epimorphisms $\gamma_1: M_1 \twoheadrightarrow S$, $\gamma_2: M_2 \twoheadrightarrow S$. Let $f_0 = (\gamma_1, \gamma_2)$: $M_1 \oplus M_2 \twoheadrightarrow S$. If $f = (\alpha, \beta): M_1 \oplus M_2 \twoheadrightarrow S$ is an epimorphism one of α and β, say α, must be an epimorphism since S is simple. By Lemma 9.6 below, if $\delta: M_2 \twoheadrightarrow S$ is any map we can write $\delta = \alpha\varphi$ where $\varphi: M_2 \to M_1$ i.e.,

$$\begin{array}{ccc} M_1 & \xleftarrow{\ \varphi\ } & M_2 \\ & {\scriptstyle\alpha}\searrow \quad \swarrow{\scriptstyle\delta} & \\ & S & \end{array}$$

commutes. This is clearly possible locally since $y \in P$ implies $S_y = 0$ while $y \notin P$ implies that Λ_y is maximal and hence M_{2y} is Λ_y-projective. Let $\theta: M \approx M$ be given by

$$\begin{array}{ccc} M_1 & \oplus & M_2 \\ {\scriptstyle 1}\downarrow & \swarrow{\scriptstyle\varphi} & \downarrow{\scriptstyle 1} \\ M_1 & \oplus & M_2 \end{array}$$

. Then $f\theta = (\alpha, \beta + \alpha\varphi) = (\alpha, \beta + \delta)$.

Therefore we can replace β with γ_2 using $\delta = \gamma_2 - \beta$. Since γ_2 is an epimorphism we can apply the same argument again to replace α by γ_1, i.e., we get $f\theta\theta_1 = (\gamma_1, \gamma_2) = f_0$. Clearly the same argument works if β and not α is an epimorphism.

Lemma 9.6. Let $A \xrightarrow{f} C \xleftarrow{g} B$ with A finitely presented. If for each y, there is a map φ_y making the diagram

$$\begin{array}{ccc} A_y & \xrightarrow{\ \varphi_y\ } & B_y \\ & {\scriptstyle f_y}\searrow \quad \swarrow{\scriptstyle g_y} & \\ & C_y & \end{array}$$

commute, then there is a map φ making

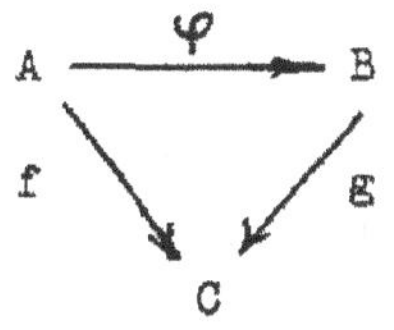

commute.

In other words, the diagram can be filled in if it can be filled in locally.

<u>Proof</u>. We must show f lies in the image of $\mathrm{Hom}_\Lambda(A, B) \xrightarrow{(1,g)} \mathrm{Hom}_\Lambda(A, C)$. Since A is finitely presented, $\mathrm{Hom}_\Lambda(A, -)$ localizes. Let $P = \mathrm{Hom}_\Lambda(A, C)$ and let Q be the image of (1, g) in P. Then for each y, $f/1 \in Q_y \subset P_y$. Write $f/1 = q/s$. Then $t(sf - q) = 0$ so $ts\, f \in Q$ for some $ts \notin y$. If $\mathfrak{A} = \{r \in R \mid rf \in Q\}$ then $\mathfrak{A}$ is an ideal and $\mathfrak{A} \not\subset y$ for all y. Thus $\mathfrak{A} = R$ so $1 \in \mathfrak{A}$ and $f \in Q$.

Combining Proposition 9.4 and 9.5 we get the following result which generalizes Bass' theorem for the case of orders. A closely related result was proved by Dress [DR] for much more general algebras.

<u>Theorem 9.7</u>. Let M be a finitely generated torsion free Λ-module such that $M = M_1 \oplus M_2$ where M_1 and M_2 are faithful Λ-modules. Let X be a finitely generated Λ-module which is a direct summand of M^n for some n. Then $X \oplus M \approx X \oplus N$ implies $M \approx N$.

<u>Proof</u>. If $M' \sim M$, then by Corollary 6.13 we have $M' = M_1' \oplus M_2'$ where $M_i' \sim M_i$ and hence M_i' is faithful. By Proposition 9.6 we see that the hypothesis of Proposition 9.5 is satisfied.

In the case considered by Bass, Λ^2 is a local direct summand of M. Thus $M = \Lambda \oplus M_1$ by Theorem 9.1 and M_1 is faithful since $\Lambda \subset M_1$ locally. Also X is assumed projective so X is a direct summand of Λ^n and hence of M^n.

We now turn to Jacobinski's theorem. For this, K must be a global field, i.e., an algebraic number field or a function field of dimension 1 over a finite field. If A is a central simple algebra over K, we say that a prime (= valuation) of K is unramified in A if the completion $\hat{A}_y$ is isomorphic to $M_n(\hat{K}_y)$. Otherwise we say y is ramified. A real archimedean prime y is ramified in A if and only if $\hat{A}_y \cong M_m(\mathbb{H})$ where $\mathbb{H}$ is the algebra of quaternions.

If R is a Dedekind ring with quotient field K, we say a prime of K comes from R if it is non-archimedian and its valuation ring is R_p for some prime ideal P of R.

__Definition__. Let K be a global field and let R be a Dedekind ring with quotient field R. Let A be a simple separable K-algebra with center C and let R' be the integral closure of R in C. We say A satisfies Eichler's condition (relative to R) if either

(1) There is a prime of K which is unramified in A and which does not come from R, or

(2) char K = 0 and $\dim_C A \neq 4$.

If A is a semisimple separable K-algebra then $A = A_1 \times \dots \times A_s$ where the A_i are simple. We say A satisfies Eichler's condition if and only if all A_i do.

Remark. Suppose K is an algebraic number field and R is the ring of integers of K. Then A satisfies Eichler's condition if and only if no A_i is a totally definite quaternion algebra. (A totally definite quaternion algebra is a simple algebra such that $\hat{A}_y \cong \mathbb{H}$ for every archimedian prime of the center of A).

Definition. If R and K are as above, A is a semisimple separable K-algebra, Λ is an R-order in A, and M is a finitely generated torsion free Λ-module, we say that M satisfies Eichler's condition if the K-algebra $\mathrm{End}_K(KM)$ does.

If $\mathfrak{A}$ is an ideal of R and f, g: $M \to N$ we write $f \equiv g \bmod \mathfrak{A}$ if f - g sends M into $\mathfrak{A} N \subset N$. If Θ is an automorphism of M and $\Theta \equiv 1 \bmod \mathfrak{A}$, then also $\Theta^{-1} \equiv 1 \bmod \mathfrak{A}$ because $\Theta(\Theta^{-1} - 1)M = (1 - \Theta)M \subset \mathfrak{A} M$ so $(\Theta^{-1} - 1)M \subset \Theta^{-1}(\mathfrak{A} M) = \mathfrak{A}\Theta^{-1}(M) = \mathfrak{A} M$.

We now state the following generalized version of Eichler's theorem.

Theorem 9.8. Assume that K is a global field. Let M be a finitely generated torsion free Λ module which satisfies Eichler's condition. Then there is a finite set of primes P with the following property: If S is a simple Λ-module with $yS = 0$ for some $y \notin P$ and f, g: $M \to S$ are epimorphisms, then either $g = \alpha f$ for some automorphism α of S or there is an automorphism Θ of M with $g = f\Theta$. Furthermore, in the second case, if $yS = 0$ and $\mathfrak{A}$ is any ideal of R prime to y, we may choose Θ so that $\Theta \equiv 1 \bmod \mathfrak{A}$.

As a consequence of this we obtain Jacobinski's cancellation theorem. (The last part of Theorem 9.8 is not needed for this.)

Theorem 9.9 (Jacobinski). Assume that K is a global field. Let M be a finitely generated torsion free Λ-module which satisfies Eichler's condition. Let X be a finitely generated Λ module which is a local direct summand of M^n for some n. Then $X \oplus M \approx X \oplus N$ implies $M \approx N$.

Proof. This is an immediate consequence of Proposition 9.5 and Theorem 9.8. There are only a finite number of $M' \sim M$ by the Jordan Zassenhaus theorem. For each of these M', Theorem 9.8 gives us some P. If we take the union of all these we get a finite set which will do for all $M' \sim M$.

As an application of this, let R be the ring of integers of an algebraic number field K and let π be a finite group. If P is an indecomposable finitely generated $R\pi$-module, then $KP \approx K\pi$. Therefore if $K\pi$ has no simple component which is a totally definite quaternion algebra, Theorem 9.9 shows that $F \oplus P \approx F \oplus Q$ implies $P \approx Q$ for finitely generated projective $R\pi$-modules. This will certainly be so if K is not totally real. If K is totally real and $K\pi$ has a bad component, we will have $\mathbb{R}\pi = \mathbb{H} \times \dots$. The image of π in $\mathbb{H}$ will be a finite subgroup of $\mathbb{H}$ which spans $\mathbb{H}$ over $\mathbb{R}$. The only finite subgroups of $\mathbb{H}$ are the cyclic groups, the generalized quaternion groups Q_n and 3 exceptional groups

$\tilde{T}$, $\tilde{O}$, $\tilde{I}$ of order 24, 48, and 120. The cyclic groups cannot span $\mathbb{H}$ since $\mathbb{H}$ is non commutative. Therefore if Π has no quotient isomorphic to Q_n, $\tilde{T}$, $\tilde{O}$, or $\tilde{I}$ then cancellation holds for finitely generated $R\Pi$-modules. This will be the case if Π is abelian or simple, or of odd order, etc.

We now turn to the proof of Theorem 9.8. Our first object is to reduce to the case of a maximal order in a central simple algebra. This is the case treated by Eichler. The reduction does not depend on the fact that K is a global field so we state it a bit more generally.

Lemma 9.10. Let $\mathcal{R}$ be a class of Dedekind rings such that if $R \in \mathcal{R}$ has quotient field K, L/K is a finite separable extension and R' is the integral closure of R in L, then $R' \in \mathcal{R}$.

If the conclusion of Theorem 9.7 holds whenever $R \in \mathcal{R}$, A is central simple over K, and Λ is a maximal order, then it holds for any order in a separable semisimple algebra over K for any $R \in \mathcal{R}$.

Proof. We first reduce to the case of a maximal order. Let $\Gamma \supset \Lambda$ be maximal. Assume the theorem holds for Γ. We will show that it also holds for Λ. Choose $r \in R$, $r \neq 0$ so $r\Gamma \subset \Lambda$. We can assume all y with $r \in y$ lie in P and that P works for Γ and $\Gamma M \subset KM$. Since $K\Gamma M = KM$, ΓM satisfies Eichler's condition if M does. Let S be a simple Λ module with $S_y = 0$ for $y \in P$. There is a unique y with $yS = 0$ (i.e., $S_y \neq 0$) and $y \notin P$. We can regard

S as a Λ_y-module. But $\Lambda_y = \Gamma_y$ since $y \notin P$ so S can also be regarded as a Γ-module. If $f: M \twoheadrightarrow S$ is an epimorphism, we can factor f as $M \twoheadrightarrow M_y \twoheadrightarrow S_y = S$. Since $\Gamma_y = \Lambda_y$, we have $(\Gamma M)_y = M_y$ and $\Gamma M \twoheadrightarrow (\Gamma M)_y = M_y \twoheadrightarrow S_y \twoheadrightarrow S$ extends f to a Γ epimorphism $f': \Gamma M \twoheadrightarrow S$. In fact the extension is unique since $\mathrm{Hom}_{\Gamma}(\Gamma M, S) \twoheadrightarrow \mathrm{Hom}_{\Lambda}(M, S)$ (by restriction) is an isomorphism (check it locally).

Let $f, g: M \twoheadrightarrow S$ be epimorphisms. Extend them to $f', g': \Gamma M \twoheadrightarrow S$. If $g' = \alpha f'$ for $\alpha \in \mathrm{Aut}\, S$, then $g = \alpha f$. If not, let $\mathfrak{A}$ be any given ideal of R prime to y. Then $r\mathfrak{A}$ is also prime to y. By the assumption on Γ, there is a Γ-automorphism Θ of ΓM with $g' = f'\Theta$ and $\Theta \equiv 1 \bmod r\mathfrak{A}$ (and so $\Theta^{-1} \equiv 1 \bmod r\mathfrak{A}$). We claim $\Theta(M) \subset M$. In fact, if $x \in M$, then $\Theta(x) - x \in r\mathfrak{A}\Gamma M \subset \mathfrak{A}\Lambda M = \mathfrak{A} M$ since $r\Gamma \subset \Lambda$. Thus $\Theta(M) \subset M$ and $\Theta|M \equiv 1 \bmod \mathfrak{A}$. The same argument applies to Θ^{-1} so $\Theta^{-1}(M) \subset M$. Thus $\Theta|M$ is an automorphism of M. It clearly has the required properties.

We now consider the case where Λ is a maximal order. Let $A = A_1 \times \dots \times A_s$ with all A_i simple. Then $\Lambda = \Lambda_1 \times \dots \times \Lambda_s$ and $S = S_1 \times \dots \times S_s$, $M = M_1 \times \dots \times M_s$. Since S is simple, only one $S_i \neq 0$ and we are looking at epimorphisms $M_i \twoheadrightarrow S_i$. If Θ is an automorphism of M_i with the required properties, use $1 \times \dots \times 1 \times \Theta \times 1 \times \dots \times 1$ on M. Therefore we can assume that A is simple and Λ is maximal. Let L be the center of A and let R' be be the integral closure of R in L. As usual, considering $\Lambda \subset R'\Lambda$ we see that $\Lambda = R'\Lambda$ so Λ is an R' order (using the maximality of Λ and the separability of L over K). By our hypothesis $R' \in \mathcal{R}$ and the theorem holds for Λ as an R'-algebra. Let P' be the required set of primes of R'. Set $P = \{R \cap y' | y' \in P'\}$. If S is a simple Λ module with $S_y = 0$ for $y \in P$, then $S = yS$ for $y \in P$ so $S = y'S$ for $y' \in P'$. Let f, g: $M \twoheadrightarrow S$ be Λ-epimorphisms $g \neq \alpha f$ and let $\mathfrak{a} \subset R$ prime to y where $yS = 0$. There is a unique $y' \subset R'$ with $y'S = 0$. Clearly $y = R \cap y'$. Since $y + \mathfrak{a} = R$, we have $y' + R'\mathfrak{a} = R'$. By the hypothesis, we can find an automorphism θ of M with $g = f\theta$, $\theta \equiv 1 \bmod R'\mathfrak{a}$. Thus $(\theta - 1)M \subset R'\mathfrak{a} M = \mathfrak{a} M$ so $\theta \equiv 1 \bmod \mathfrak{a}$. Of course we must check that M satisfies Eichler's condition with respect to R' and Λ but this is obvious.

We next show that Theorem 9.8 is a consequence of the following result.

Proposition 9.11. Let R be a Dedekind ring whose quotient field K is a global field. Let A be a simple K-algebra satisfying Eichler's condition. Let Λ be an order in A over R. Then there is a finite set of primes P of R such that if $\mathfrak{b}$ is any non-zero ideal of R prime to all $y \in P$ and $f(x) = x^n + a_1 x + \ldots + (-1)^n \in R[x]$ where $\dim_K A = n^2$, then there is a unit $\eta \in \Lambda$ with $f(\eta) \equiv 0 \bmod \mathfrak{b}$. Furthermore, if $\mathfrak{a}$ is any non-zero ideal of R prime to $\mathfrak{b}$, we can choose η with $\eta \equiv 1 \bmod \mathfrak{a}$.

The following well known result about maximal orders will be useful. Let R be a Dedekind ring with quotient field K and let A be a central simple K-algebra. If Λ, Γ are orders in A we say Λ and Γ are conjugate if $\Gamma = a\Lambda a^{-1}$ for some unit a of A.

Proposition 9.12. If R satisfies the Jordan Zassenhaus theorem there are only a finite number of conjugacy classes of maximal orders in A. If R is local, there is only one class.

Proof. Fix one maximal order Λ. If Γ is any other one, let $M = \Gamma\Lambda$. This is a left Γ module and right Λ-module. By Theorem 5.5, $MM^{-1} = \Gamma$. If Γ' is another maximal order and $M' = \Gamma'\Lambda$, then $M'M'^{-1} = \Gamma'$. If $M' \approx M$ as a right Λ module, choose an isomorphism $\varphi: M \to M'$. Apply $K \otimes_R -$ and get $K \otimes_R \varphi : A \approx A$ as a right A-module. Thus $K \otimes_R \varphi : x \mapsto ax$ for all $x \in A$ where $a \in U(A)$. Therefore $M' = \varphi(M) = aM$ so $\Gamma' = aM(aM)^{-1} = aMM^{-1}a^{-1} = a\Gamma a^{-1}$. Thus the number of conjugacy classes of maximal orders is $\leq$ the number of isomorphism classes of right Λ-ideals. If R satisfies the Jordan Zassenhaus theorem this is finite. If R is local, there is only one class by Theorem 5.27.

Corollary 9.13. If A is unramified at $y \subset R$, then $\Lambda/y\Lambda = M_n(R/y)$ where $n^2 = \dim_K A$.

Proof. Since y is unramified, $\hat{A}_y \approx M_n(\hat{K}_y)$. By Theorem 5.28 and Lemma 7.2, $\hat{\Lambda}_y$ is a maximal order in $\hat{A}_y$. By Lemma 7.4, so is $M_n(\hat{R}_y)$ so $\hat{\Lambda}_y \approx M_n(\hat{R}_y)$ by Proposition 9.12. Thus $\Lambda/y\Lambda \approx \hat{\Lambda}_y/y\hat{\Lambda}_y \approx M_n(\hat{R}_y/y\hat{R}_y) \approx M_n(R/y)$.

The following result will not be needed here but we include it for completeness.

Corollary 9.14. Let Λ be a maximal order and M a finitely generated torsion free Λ-module. Then $\Gamma = \mathrm{End}_{\Lambda}(M)$ is a maximal order in $B = \mathrm{End}_A(KM)$.

Proof. Let $A = A_1 \times \ldots \times A_s$. Then $\Lambda = \Lambda_1 \times \ldots \times \Lambda_s$, $M = M_1 \times \ldots \times M_s$ and we reduce immediately to the case where M is simple. If R' is the integral closure of R in the center of A (= center of B), then maximal orders over R and R' are the same so we can assume A (and hence B) is central simple. By Corollary 5.29 we can assume R is a DVR. Let $A = M_r(D)$ where D is a division ring. Let Σ be a maximal order of D. Then $M_r(\Sigma)$ is a maximal order of A by Lemma 7.4. By Proposition 9.12, $\Lambda \approx M_r(\Sigma)$ so we can assume $\Lambda = M_r(\Sigma)$. Let $N = \coprod_1^r \Sigma$ as a left Λ-module. Since KN is a simple A-module, Theorem 5.27 shows that $M \approx \coprod_1^s N$ for some s. Thus $\Gamma = \mathrm{End}_{\Lambda}(M) \approx M_s(\mathrm{End}_{\Lambda} N)$. But $\mathrm{End}_{\Lambda}(N) \approx \Sigma^0$ which is a maximal order. By Lemma 7.4 so is $M_s(\mathrm{End}_{\Lambda}(N))$.

To prove Theorem 9.8 from Proposition 9.11, we require the following lemma.

Lemma 9.15. Let k be a field, S a simple $M_m(k)$-module, M a finitely generated $M_m(k)$-module. Suppose there is no automorphism α of S with $h = \alpha g$. Let f(x) be a polynomial over k with no root in k. Let $\Theta: M \to M$ be an automorphism of M with $f(\Theta) = 0$. Then Θ is an automorphism of M and there is an automorphism τ of M such that $g \cdot \tau \Theta \tau^{-1} = h$.

<u>Proof</u>. Let $M = \coprod_{1}^{r} S$. Then $\mathrm{End}(M) = M_r(\mathrm{End}\ S) = M_r(k)$. This acts from the right on $\mathrm{Hom}(M, S) = k^r = V$ say. Now $\Theta \in \mathrm{End}(M) = M_r(k)$ and $g, h \in V$. Note that g and h are linearly independent over k since $h \neq \alpha g$ for any $\alpha \in \mathrm{End}(S) = k$. Let $f(x) = a_r x^r + \dots + a_0$. Since f has no root in k, $a_0 = f(0) \neq 0$. Thus $-a_0^{-1}(a_r \Theta^{r-1} + \dots + a_1)$ is an inverse for Θ. We claim there is an e in V such that e and $e\Theta$ are linearly independent over k. If not, then for each $v \in V$, $v\Theta = \lambda v$ with $\lambda \in k$. If $v_1, v_2 \in V$ and $v = v_1 + v_2$ we have $v_1\Theta = \lambda_1 v_1$, $v_1\Theta = \lambda_2 v_2$, $v\Theta = \lambda_1 v_1 + \lambda_2 v_2 = \lambda(v_1 + v_2)$. Thus $\lambda_1 = \lambda_2$ so $\Theta = \lambda I$. Since $f(\Theta) = 0$, we have $f(\lambda) = 0$ but this is impossible since $\lambda \in k$. Thus e must exist. Let τ^{-1} be an automorphism of V over k sending e to g and $e\Theta$ to h. Then $g\tau\Theta\tau^{-1} = e\Theta\tau^{-1} = h$.

We now prove Theorem 9.8 for an order Λ in a central simple K-algebra A.

Let P be a finite set of primes of R including all those given by Proposition 9.11 and all those ramified in A. If $y \notin P$, $\Lambda/y\Lambda \cong M_r(R/y)$ by Corollary 9.13. Now suppose M is a torsion free finitely generated Λ-module, S is a simple Λ-module, and g, h: $M \twoheadrightarrow S$ are epimorphisms such that $h \neq \alpha g$ for any automorphism α of S. Suppose $yS = 0$ for $y \notin P$. Then g, h factor as $M \twoheadrightarrow M/yM = \bar{M} \rightrightarrows S$. We will apply Proposition 9.12 to the order $\Gamma = \mathrm{End}_\Lambda(M) \subset B = \mathrm{End}_A(KM)$. Clearly B is central simple over K. Let $\dim_K B = m^2$. By Lemma 9.16 below we can find a polynomial $\bar{f}$ over $k = R/y$ of the form $\bar{f}(x) = x^m + \bar{a}_1 x^{m-1} + \ldots + (-1)^m$ with no root in k. Lift $\bar{f}$ to $f(x) = x^m + a_1 x^{m-1} + \ldots + (-1)^m \in R[x]$. Applying Proposition 9.11 we get a unit $\eta \in \Gamma$ with $\bar{f}(\eta) \equiv 0 \bmod y$ and with $\eta \equiv 1 \bmod \mathfrak{c}$ where $\mathfrak{c}$ is any chosen ideal of R prime to y. Let $\bar{\eta}$ be the automorphism of $\bar{M} = M/yM$ induced by η. Then $\bar{f}(\bar{\eta}) = 0$. Lemma 9.15 now shows that $\bar{h} = \bar{g}\tau\bar{\eta}\tau^{-1}$ for some automorphism τ of $\bar{M}$. We claim that if τ is properly chosen, we can lift $\tau\bar{\eta}\tau^{-1}$ to an automorphism θ of M.

There are only a finite number of $\tau \in \mathrm{Aut}\,\bar{M}$. Since M is projective over Λ, we can lift each $\tau_i \in \mathrm{Aut}\,\bar{M}$ to an endomorphism $\varphi_i \in \mathrm{End}_\Lambda(M)$. Since $\varphi_i : M \to M$ induces $\tau_i : M/yM \approx M/yM$, Nakayama's lemma shows that $\varphi_i : M_y \twoheadrightarrow M_y$ is an epimorphism and hence an isomorphism. Thus $\varphi_i^{-1} : M_y \to M_y$ is defined. Since M and $M_i = \varphi_i^{-1}M$ lie in M_y we can find some $r \in R - y$ such that $rM_i \subset M$ for all i. Let $\mathfrak{a}$ be the ideal chosen in Theorem 9.8. Since $\mathfrak{a}$ is prime to y we have $\mathfrak{a} \not\subset y$. Choose $a \in \mathfrak{a}$ with $a \notin y$. We now let $\mathfrak{c} = (ra)$ so that $\eta = 1 + ra\psi$ where $\psi : M \to M$. Since $\eta^{-1} - 1 = (\eta - 1)\eta^{-1}$. We can also write $\eta^{-1} = 1 + ra\Psi$. Now, considering all maps as endomorphisms of M_y, we have $\Theta = \varphi_i \eta \varphi_i^{-1} = 1 + \varphi_i a \Psi(r\varphi_i^{-1})$. Thus $\Theta\colon M \to M$ and $\Theta \equiv 1 \bmod (a)$. The same argument applies to $\Theta^{-1} = \varphi_i \eta^{-1} \varphi_i^{-1}$ so $\Theta^{-1}\colon M \to M$. Thus Θ is an automorphism of M. Since $\Theta\varphi_i = \varphi_i \eta$, reducing mod y shows that $\bar{\Theta}\tau_i = \tau_i \bar{\eta}$ or $\bar{\Theta} = \tau_i \bar{\eta} \tau_i^{-1}$. Thus $\bar{h} = \bar{g}\bar{\Theta}$ so $h = g\Theta$.

Lemma 9.16.. If k is any finite field and $m \geqslant 1$ is an integer, then there is a polynomial $f(x) = x^m + a_1 x^{n-1} + \ldots + (-1)^m \in k[x]$ with no root in k.

Proof. Let $q = |k|$. The number of f(x) of the required form is q^{m-1}. If f has a root $\gamma \in k$, then $f(x) = (x - \gamma)(x^{m-1} + b_1 x^{m-2} + \ldots + b_{m-1})$ where $-\gamma b_{m-1} = (-1)^m$. Thus $b_{m-1} \neq 0$ and b_{m-1} determines γ. The number of possibilities for such f(x) is $\leq q^{m-2}(q - 1) < q^{m-1}$ so some f(x) does not have this form.

We must now prove Proposition 9.11. We begin by recalling some classical results which will be needed in the proof.

Let A be a central simple K-algebra. We say A splits if $A \approx M_n(K)$. If y is a prime (i.e., a valuation) of K we say that A splits at y if the completion $\hat{A}_y = \hat{K}_y \otimes_K A$ splits (i.e., is $\approx M_n(\hat{K}_y)$). Thus y is ramified in A if and only if $\hat{A}_y$ does not split. Eichler's method depends on the following classical result. It is here that it is essential to assume K is a global field.

Theorem 9.17 (Albert, Brauer, Hasse, Noether). Let A be a central simple algebra over K.

(1) If K is a global field, there are only a finite number of primes of K ramified in A.

(2) If K is a global field and no primes of K ramify in A, then A splits.

(3) If K is a local field and $[L: K] = n$ where $\dim_K A = n^2$ then L splits A, i.e., $L \otimes_K A$ splits.

The deepest part of this is (2) which says that an algebra which splits locally also splits globally.

Proof. We shall only indicate the relation between the theorem and cohomological results of class field theory. We use the notation of AT. Let E be a galois extension of K which splits A. Choosing a factor set for A with respect to E gives a cohomology class $\xi \in H^2(G, E^*)$ where G is the galois group of E/K, and A splits if and only if $\xi = 0$ (see CL X § 5). Consider $0 \to E^* \to J_E \to C_E \to 0$. This gives

$$H^1(G, C_E) \xrightarrow{\delta} H^2(G, E^*) \longrightarrow H^2(G, J_E) \longrightarrow H^2(G, C_E)$$

Now $H^2(G, J_E) = \coprod_y H^2(G_P, E_P^*)$ where y runs over the primes of K, E_P is the completion of E at any chosen P over y, and $G_P = G_{E_P/K_y}$. The image of ξ in this is (ξ_y) where ξ_y is the cohomology class of $\hat{A}_y$. Thus $\hat{A}_y$ splits if and only if $\xi_y = 0$. This is true for almost all y since we are looking at a direct sum. This proves (1). If all $\xi_y = 0$, then ξ comes from an element of $H'(G, C_E)$ but this is 0 by global class field theory so $\xi = 0$. This proves (2). For (3) we use local class field theory which gives us a monomorphism inv: $H^2(G, E^*) \to \mathbb{Q}/\mathbb{Z}$. Choose $E \supset L \supset K$. Then we have

$$\begin{array}{ccc} H^2(G_{E/K}, E^*) & \xrightarrow{\text{res}} & H^2(G_{E/L}, E^*) \\ \downarrow & & \downarrow \\ \mathbb{Q}/\mathbb{Z} & \xrightarrow{[L:\,K]} & \mathbb{Q}/\mathbb{Z} \end{array}$$

The top map sends ξ to the cohomology class ξ' of $L \otimes_K A$. Since the vertical maps are monomorphisms we see that if $\xi' = 0$ for one extension L/K of some degree d then $\xi' = 0$ for every L/K with $d/[L:\,K]$. The existence of a splitting field of degree dividing n follows from the next theorem. This is also classical (but elementary) and will be needed in the proof of Eichler's theorem.

Theorem 9.18. Let A be a central simple algebra over a field K. Then $A = M_r(D)$ where D is a division ring with center K. Let $\dim_K A = n^2$ and $\dim_K D = s^2$ so $n = rs$. Let E be an extension field of K of degree $e = [E\, K]$. Then

(1) If E splits A, i.e., $E \otimes_K A \cong M_n(E)$, then $s \mid e$.

(2) If $e = n$ then E splits A if and only if E is isomorphic as a K-algebra to a subalgebra of A (we will write $E \subset_K A$ for this).

(3) If $E \subset_K A$ then $e \mid n$.

(4) If $E \subset_K A$ is a maximal commutative subfield of A, then E splits A.

<u>Remark</u>. If A = D it follows that every maximal commutative subfield of D has degree n. This is not so in general. For example, if $A = M_n(\mathbb{C})$ only e = 1 is possible.

<u>Proof</u>. By Wedderburn's theory D is uniquely determined by A (as $\mathrm{End}_A(S)^0$ where S is a simple module). The same result for $E \otimes_K A$ shows that E splits A if and only if E splits D.

Suppose first that E splits A. Then $D \otimes E \approx M_s(E)$. Let S be a simple module over $D \otimes E$. Then S is a D, E-bimodule. Since S is a simple $M_s(E)$-module, it is isomorphic to the module s rowed column vectors. Thus $\dim_E S = s$ so $\dim_K S = se$. Now regard S as a D-module. Let $t = \dim_D S$. Then $se = \dim_K S = t \circ \dim_K D = ts^2$ so $e = ts$. This proves (1). Since S is a D, E-bimodule, we get $E \to \mathrm{End}_D S = \mathrm{End}_D(D^t) = M_t(D)^o$. This is injective since E is a field. If e = n then t = e/s = r so $M_t(D) \approx A$. Therefore $E \hookrightarrow A^o$ so $E = E^o \hookrightarrow A$. This proves half of (2).

Now suppose $E \subset A$. Let $S = D^r = \{(d_1, \ldots, d_r)\}$. This is a left D-module and a right A-module. Since $E \subset A$, we can regard S as an E module and thus as a $D \otimes_K E$-module. Since E is commutative we get a ring homomorphism $\varphi: D \otimes_K E \to \mathrm{End}_E(S) = M_u(E)$ where $u = \dim_E S$. This map is injective since $D \otimes_K E$ is simple. Now $\dim_K S = ue$ but $\dim_K S = \dim_K D \cdot \dim_D S = rs^2 = ns$. If e = n, then $u = s$ and $\dim_E D \otimes_K E = s^2 = \dim_E M_u(E)$. Therefore φ is an isomorphism so E splits D. This proves the other half of (2). If we do not assume $e = n$ we have $D \otimes_K E \approx M_h(D')$ for some division ring D' with center E. Let $\dim_E D' = v^2$. A simple module over $M_h(D')$ has dimension hv^2 over E and therefore has dimension ehv^2 over K. Since S is a module over $M_h(D')$, $ehv^2 | \dim_K S = r \dim_K D = rs^2$. But $h^2v^2 = \dim_E M_h(D') = \dim_E(D \otimes_K E) = s^2$ so $ehv^2 | rh^2v^2$ or $e | rh$. Since $hv = s$ we get $e | rs = n$. This proves (3). Finally suppose E is maximal in A. We have $E \to \mathrm{End}_{D \otimes E}(S) \subset \mathrm{End}_D(S) = A^0$. This clearly gives the original embedding of E in A by the definition of the action of E on S. Since $D \otimes E = M_h(D')$ we see that $\mathrm{End}_{D \otimes E}(S) = M_m(D')^0$ for some m. Clearly $E \to \mathrm{End}_{D \otimes E}(S)$ is just the embedding of E as the center of $\mathrm{End}_{D \otimes E}(S)$. Therefore we get $E \subset M_m(D') \subset A$. If $D' \neq E$ we can adjoin an element of D' to E and get a larger commutative subfield of A. Since E is maximal, $D' = E$. Therefore, $D \otimes_K E = M_h(E)$ so E splits D. This proves (4).

Combining the two preceding theorems we get the following result which is the main tool used in Eichler's method.

<u>Proposition 9.19</u>. Let K be a global field and A a central simple algebra over K with $\dim_K A = n^2$. Let $f(x) = x^n + a_1x^{n-1} + \ldots + a_n$ be a monic polynomial over K. Assume

(1) If y is a mon archimedian prime of K which is ramified in A, then f(x) is irreducible over $\hat{K}_y$.

(2) If y is an archimedian prime of K which is ramified in A then f(x) has no root in $\hat{K}_y$.

(3) f(x) is irreducible over K.

Then there is an $\alpha \in A$ with $f(\alpha) = 0$ and f(x) is the reduced characteristic equation of α.

Note that (3) is satisfied if the set of primes considered in (1) is non-empty.

Proof. Let $E = K(\alpha)$ where α is a root of f. Then $[E: K] = n$ by (3). If y is one of the primes considered in (1) and P is any extension of y to E, then $\hat{E}_P = \hat{K}_y(\alpha)$ so $[\hat{E}_P: \hat{K}_y] = n$. By Theorem 9.17 (3), $\hat{E}_P \otimes_{\hat{K}_y} \hat{A}_y = \hat{E}_P \otimes_E (E \otimes_K A)$ splits. This is also true for y unramified in A since $\hat{A}_y$ is then already split. If y is archimedian and ramified, then $\hat{K}_y \approx \mathbb{R}$ but $\hat{E}_P \neq \hat{K}_y$ by (2). Thus $\hat{E}_P = \mathbb{C}$. Since this is algebraically closed, $\hat{E} \otimes_E (E \otimes_K A)$ splits. By Theorem 9.17 (2), we conclude that $E \otimes_K A$ splits. Since $[E: K] = n$, Theorem 9.18 (2) shows that $E \subset_K A$. Regarding this as an inclusion we see that $\alpha \in E \subset A$ but $f(x) = 0$. Finally, let g(x) be the reduced characteristic equation of α. Then $g(\alpha) = 0$ by the Cayley-Hamilton theorem applied to $L \otimes_K A \approx M_n(L)$ for any splitting field L. By (3), $f(x) | g(x)$ but f and g are both monic of degree n so $f = g$.

In order to apply this we must construct the polynomial f. We do this by examining each prime separately, finding a polynomial

$f_y(x) = x^n + \dots$ over K_y and applying the approximation theorem 9.21 to get f. The following lemma shows that this will work. Recall that a valuation on K makes K a metric space with $d(x, y) = |x - y|$. If we have two polynomials $f(x) = a_0x^n + a_1x^{n-1} + \dots + a_n$ and $g(x) = b_0x^n + b_1x^{n-1} + \dots + b_n$ we define $|f - g| = \max|a_i - b_i|$.

Lemma 9.20. Let K be a local field and let $f(x) = x^n + a_1x^{n-1} + \dots + a_n$ and $g(x) = x^n + b_1x^{n-1} + \dots + b_n$ be monic polynomials over K of fixed degree n.

(1) If $K \neq \mathbb{R}$ or $\mathbb{C}$ and f(x) is irreducible over K, then there is an $\varepsilon > 0$ so that $|f - g| < \varepsilon$ implies that g is also irreducible over K. Similarly if f has distinct roots there is an $\varepsilon > 0$ so $|f - g| < \varepsilon$ implies that g has distinct roots.

(2) If $K = \mathbb{R}$ and f has no root in K, there is an $\varepsilon > 0$ so that $|f - g| < \varepsilon$ implies that g has no roots in K.

Proof. We can prove (1) using Hensel's lemma. Instead, following Eichler we give a proof using the fact that for any integer m, $\{x \in K | \operatorname{ord} x \geq m\}$ is compact. Let $\operatorname{ord}(c_0x^m + \dots + c_m) = \min \operatorname{ord} c_i$ (this agrees with the definition of $|f| = |f - 0|$ above). Then $\operatorname{ord}(p(x)q(x)) = \operatorname{ord} p(x) + \operatorname{ord} q(x)$ by Gauss' lemma (to prove this multiply by constants so $\operatorname{ord} p = \operatorname{ord} q = 0$. Thus $p, q \in \mathcal{O}[x]$ where $\mathcal{O}$ is the valuation ring and $p, q \not\equiv 0 \bmod y$. Taking images in $\mathcal{O}/y[x]$ we have $\overline{pq} = \bar{p}\bar{q} \neq 0$ so $pq \not\equiv 0 \bmod y$). Suppose now that there are monic polynomials g_v of degree n with $|f - g_v| \to 0$ but $g_v = p_v q_v$ where deg p_v and deg $q_v < n$. We can assume p_v and q_v are monic so ord p_v, ord $q_v \leq$ ord $1 = 0$. Therefore ord p_v, ord $q_v \geq$ ord g_v = ord f for large v so the coefficients lie in a compact set. Taking subsequences we can assume that p_v and q_v have constant degrees and that $p_v \to p$, $q_v \to q$. Since $g_v = p_v q_v \to f$ we get $f = pq$, a contradiction.

For the second part of (1) note that f has distinct roots if and only if $\gcd(f, f') = 1$. Suppose we have $g_v \to f$ with $d_v = \gcd(g_v, g_v')$ of degree > 0. Assume d_v monic. Taking subsequences as above we can assume d_v has constant degree, $d_v \to d$, $g_v/d_v \to p$, $g_v'/d_v \to q$. Then $f = dp$, $f' = dq$, a contradiction.

For (2) suppose $g_v \to f$ and $g_v(c_v) = 0$ for some $c_v \in K$. If $g = x^n + b_1 x^{n-1} + \ldots + b_n$ and $g(c) = 0$, then $|c^n| \leq \sum_1^n |b_i||c|^{n-i}$ so $|c| \leq \sum_1^n |b_i||c|^{1-i}$. Since $1 - i \leq 0$ this shows that $|c| \leq \max(1, \sum |b_i|)$. Thus the c_v are bounded. Taking a subsequence, we assume that $c_v \to c$. Then $f(c) = 0$, a contradiction.

Now assuming we can find a polynomial $f_y(x)$ for each y in Proposition 9.19 we can find a single polynomial which will do by the approximation theorem and Lemma 9.20. However, to prove Eichler's theorem it is essential to have a polynomial with coefficients in R.

Definition. Let S be a set of real valued valuations on a field K. We say that S satisfies the Strong Approximation Theorem (SAT) if for any finite set $v_1, \ldots, v_n \in S$ and $x_1, \ldots, x_n \in K$ and any $\varepsilon > 0$ we can find $x \in K$ such that $|x - x_i|_{v_i} < \varepsilon$ for $i = 1, \ldots, n$ and $|x|_v \leq 1$ for all other $v \in S$, $v \neq v_1, \ldots, v_n$.

The ordinary approximation theorem says that a finite S satisfies the SAT.

Theorem 9.21. If K is a global field and S is any proper subset of the set of all primes of K, then S satisfies the SAT.

For the proof see OM 33:11 or the appendix of these notes. The restriction on S is clearly necessary.

To construct our polynomial locally, we use the following lemma.

Lemma 9.22. Let K be a local field not $\mathbb{R}$ or $\mathbb{C}$. Let $n \geq 1$ be an integer. Then there are polynomials over K of the form $x^n + a_1 x^{n-1} + \ldots + (-1)^n$ which are separable, irreducible, and arbitrarily close to $(x - 1)^n$.

Proof. Let L be the unramified extension of K of degree n. (CL III § 5, OM § 32B) Suppose we can find $\zeta \in L$ such that $L = K(\zeta)$, $N_{L/K}\zeta = 1$, and ζ is arbitrarily close to 1. Then the minimal polynomial f(x) of ζ will do. It is certainly separable and irreducible. Also $f(x) = \prod(x - \zeta_i)$ where ζ_i are the conjugates of ζ. These are all near 1 since $|\zeta_i - 1| = |\zeta - 1|$.

(The valuation on L extending that of K is unique and hence invariant under the galois group.) Thus $f(x)$ is near $(x - 1)^n$. Also $f(0) = (-1)^n N_{L/K} \zeta = (-1)^n$.

If n is prime to the characteristic of K it is very easy to find such a ζ. Let $L = K(\theta)$ and let $\pi \in K$ with ord $\pi = 1$. If h is large, $1 + \pi^h\theta$ is near 1 and so is $N_{K/L}(1 + \pi^h\theta)$. Thus we can write $N_{K/L}(1 + \pi^h\theta) = a^n$ with $a \in K$ near 1. (W Proposition 3-1-6.) Let $\zeta = a^{-1}(1 + \pi^h\theta)$. This uses only the fact that L/K is separable of degree n.

To do the general case, we use a different argument. The extension L/K is cyclic. Let σ generate its galois group G. (Any $\sigma \neq 1$ will do.) By the normal basis theorem there is some $\theta \in L$ such that the $\tau\theta$ for $\tau \in G$ form a base for L/K. Let $\alpha_h = 1 + \pi^h\theta$ and let $\zeta_h = \alpha_h^{-1}\sigma(\alpha_h)$. This is near 1 for h large and $N_{L/K}(\zeta_h) = 1$. We claim there are arbitrarily large values of h with $L = K(\zeta_h)$. If not, for each large h we can find $\tau_h \in G$ with $\tau_h(\zeta_h) = \zeta_h$ and $\tau_h \neq 1$. Pick a subsequence of h's with $\tau_h = \tau$. Since $\tau\zeta_h = \zeta_h$ we have $\tau\alpha^{-1}\cdot\tau\sigma\alpha = \alpha^{-1}\sigma\alpha$ or $\alpha\cdot\tau\sigma(\alpha) = \sigma(\alpha)\cdot\tau(\alpha)$ where we write α for α_h. This says

$$(*) \quad (1 + \pi^h\theta)(1 + \pi^h\tau\sigma(\theta)) = (1 + \pi^h\sigma(\theta))(1 + \pi^h\tau(\theta)).$$

Subtract 1 from both sides, divide by Π^h and take the limit along the chosen subsequence. We get $\theta + \tau\sigma\theta = \sigma\theta + \tau\theta$. But this is impossible since the $\rho\theta$, $\rho \in G$ form a base unless char K = 2. In fact $\sigma\theta, \tau\theta \neq \theta$ so the only term which can cancel the θ on the left is $\tau\sigma\theta$. Thus $\tau\sigma = 1$ and the left side is 2θ. If char K = 2, $2\theta = 0$ so we get $\sigma\theta + \tau\theta = 0$. Thus $\sigma = \tau$ and $\sigma^2 = 1$. The equation (*) now becomes $(1 + \Pi^h\theta)^2 = (1 + \Pi^h\sigma\theta)^2$ which implies $\theta^2 = \sigma\theta^2$ and so $\theta = \sigma\theta$, a contradiction.

We can now prove Proposition 9.11. By Proposition 9.12, we can find maximal orders $\Gamma_1, \ldots, \Gamma_h$ in A such that any maximal order is conjugate to one of the Γ_i. Choose $r \in R$, $r \neq 0$ such that $r\Gamma_i \subset \Lambda$ for $i = 1, \ldots, h$. Let P consist of the primes of R containing r and all primes of R which are ramified in A. We can also assume $P \neq \emptyset$ by throwing in any prime of R provided $R \neq K$. If $R = K$, Proposition 9.11 is trivial.

We now construct the required polynomial using Eichler's condition. Let $\mathfrak{a}$ and $\mathfrak{b}$ be the ideals given in Proposition 9.11. Choose $a \in \mathfrak{a}$, $a \neq 0$.

Lemma 9.23. Under the hypotheses of Proposition 9.11 we can find a polynomial $g(x) = x^n + a_1x^{n-1} + \ldots + (-1)^n \in R[x]$ such that

(1) For each non-archimedian prime y of K which is ramified in A, each $y \in P$, and each $y \ni a$ in R, g(x) is arbitrarily near $(x - 1)^n$ over $\hat{K}_y$.

(2) For each prime y of K which occurs in (1), g(x) is irreducible over $\hat{K}_y$.

(3) For each archimedian y which ramified in K, $g(x)$ has no zero in $\hat{K}_y$.

(4) For each $y \supset \mathfrak{b}$ in R, $g(x)$ is arbitrarily near $f(x)$.

Proof. For each y concerned we can find a polynomial $g_y(x)$ satisfying the required condition over $\hat{K}_y$ and with $g_y(0) = (-1)^n$. For the non-archimedian primes we use Lemma 9.22. If there is an archimedian prime y ramified in A, then $\hat{A}_y = M_m(\mathbb{H})$ so $n = 2m$ is even and we can let $g_y(x) = x^n + 1$. Each $g_y(x)$ has coefficients integral at y if y is non-archimedian provided we take it near enough to $(x - 1)^n$.

Suppose first that there is a prime y_0 of K which is unramified in A and which does not come from R. By Theorem 9.21 the set of $y \neq y_0$ satisfies the SAT. Therefore we can find $g(x)$ arbitrarily near $g_y(x)$ for each of the above y and such that all coefficients of $g(x)$ satisfy $|a|_y \leq 1$ for the remaining $y \neq y_0$. Therefore all coefficients of g lie in R. By Lemma 9.20, conditions (2) and (3) will be satisfied if g is close enough to g_y. Conditions (1) and (4) are clear. We can clearly choose the constant term of g to be $(-1)^n$ (and choose g monic).

Suppose now that char $K \neq 0$ and $\dim_K A \neq 4$. We then proceed as above but ignore the archimedian primes. This gives a polynomial $h(x)$ satisfying (1), (2), and (4). If some archimedian y ramifies in A, then $n = 2m$ as above and $h(x) = x^{2m} + \ldots + 1$. Let q be a positive integer highly divisible by all the non-archimedian y occuring in (1), (2), and (4). Let $g(x) = h(x) + qx^2$. If q is sufficiently divisible, then g is very near h at all y in (1), (2), and (4). If q is sufficiently large we will have $g(x) > 0$ for all $x \in \hat{K}_y$ for each real archimedian y. The condition $\dim_K A \neq 4$ means $n > 2$ so g is monic. This proves the lemma.

By taking the approximation sufficiently close we can insure that $(x - 1)^n - g(x)$ has coefficients in $r^n a^n R$ and that $f(x) - g(x)$ has coefficients in $\mathfrak{b}$. Since we have assumed $P \neq \emptyset$, g(x) is irreducible by (2). By Proposition 9.19, we can find an element $\alpha \in A$ with $g(\alpha) = 0$. Write $g(x) - (x - 1)^n = (ra)^n h(x)$ where h(x) has coefficients in R. Note that $\deg h(x) < n$. Since $g(\alpha) = 0$ we have $(\alpha - 1)^n + (ra)^n h(\alpha) = 0$ so $\alpha - 1$ is a root of the polynomial $x^n + (ra)^n h(x + 1)$. Therefore if we let $\gamma = (\alpha - 1)/ra$ we see that γ is a root of $(rax)^n + (ra)^n h(rax + 1)$ or of $x^n + h(rax + 1)$. This is monic and has coefficients in R so γ is integral over R. By Lemma 9.24 below, γ lies in some maximal order Γ of A. Let $x \in U(A)$ with $x\Gamma x^{-1} = \Gamma_i$ for some $i = 1, \ldots, h$, and set $\eta = x\alpha x^{-1}$. Then $g(\eta) = 0$, $\eta \in \Gamma_i$ and $\eta = 1 + ra\delta$ where $\delta = x\gamma x^{-1} \in \Gamma_i$. By the choice of r, $r\Gamma_i \subset \Lambda$ so $r\delta \in \Lambda$. Thus $\eta \in \Lambda$ and $\eta \equiv 1 \bmod (a)$. Since $f(x) - g(x)$ has coefficients in $\mathfrak{b}$, $f(\eta) = f(\eta) - g(\eta)$ lies in $\mathfrak{b}\Lambda$. Finally η is a unit of Λ because $(\eta^{n-1} + a_1\eta^{n-2} + \ldots + a_{n-1})\eta = (-1)^{n-1}$.

<u>Lemma 9.24</u>. Let A be a separable semisimple K-algebra. Let $\gamma \in A$ be integral over R. Then γ lies in a maximal order of A.

<u>Proof</u>. The ring $R[\alpha]$ is a finitely generated R-module. Let $w_1, \ldots, w_m$ be a base for A over K and let

$M = \sum^{m} R[\alpha]w_i$. This is a finitely generated R-module and $KM = A$. Thus $\Theta_\chi(M) = \{x \in A \mid xM \subset M\}$ is an order. It clearly contains $R[\alpha]$. Embed it in a maximal order by Proposition 5.1.

This completes the proof of Proposition 9.11 and so of Theorem 9.8 and 9.9. Eichler's condition was used only in Lemma 9.23 to insure that the coefficients of g(x) lie in R. Without this there would be no hope of getting η to lie in Λ. The following example shows that the condition is necessary.

<u>Example</u>. Let Π be the generalized quaternion group of order 32. Then there is an ideal I in $\mathbb{Z}\Pi$ with $I \oplus \mathbb{Z}\Pi \approx \mathbb{Z}\Pi \oplus \mathbb{Z}\Pi$ but $I \not\approx \mathbb{Z}\Pi$. There is a maximal order Γ of $\mathbb{Q}\Pi$ containing $\mathbb{Z}\Pi$ such that $\Gamma I \oplus \Gamma \approx \Gamma \oplus \Gamma$ but $\Gamma I \approx \Gamma$.

The proof may be found in SP.

As a further application of Proposition 9.19 we now give Eichler's proof of Theorem 7.6. He actually proves a somewhat stronger result.

<u>Theorem 9.25</u>. (1) Let K be a global field and let A be a central simple K-algebra. Then the image of the reduced norm

$\eta: U(A) \longrightarrow U(K)$ is $\{a \in U(K) \mid a > 0$ at each real archimedian prime which ramifies in A$\}$.

(2) Let R be a Dedekind ring with quotient field K. If $a \in R$ satisfies the conditions of (1) and if A satisfies Eichler's condition, then $a = n(\alpha)$ with $\alpha \in U(A)$ integral over R.

Proof. We have already observed the necessity of the condition on a in Chapter 7. Suppose we can find a polynomial $f(x) = x^n + a_1x^{n-1} + \ldots + (-1)^n a$ satisfying the conditions of Proposition 9.19. Then by that result we can find $\alpha \in A$ with $f(\alpha) = 0$ and f will be the reduced characteristic equation of α. In particular $n(\alpha) = (-1)^n f(0) = a$. In case (2) we must find such an f in R[x]. Then α will be integral over R. By Lemma 9.26 below we can find a polynomial $f_y(x)$ of the required form for each non-archimedian prime of K which is ramified in A (and one other such prime if no such primes ramify in A). If there is a real ramified prime y, then $\hat{A}_y = M_m(\mathbb{H})$ so $n = 2m$ is even and we can take $f_y(x) = x^n + a$ since $a > 0$ in $\hat{K}_y$. For (1) we need only take f(x) close enough to each $f_y(x)$ for the finite number of y considered. For (2) we proceed as in Lemma 9.23 getting $f(x) \in R[x]$.

Lemma 9.26. Let K be a local field not $\mathbb{R}$ or $\mathbb{C}$. Let $a \in K$, $a \neq 0$ and let $n > 0$ be any integer. Then there is a separable polynomial $f(x) = x^n + a_1x^{n-1} + \ldots + (-1)^n a$ which is irreducible over K. If $a \in \mathcal{O}$, the valuation ring of K, we can choose $f(x) \in \mathcal{O}[x]$.

Proof. As in the proof of Lemma 9.22 it will suffice to find a separable extension L/K of degree n and an $\alpha \in L$ so that $L = K(\alpha)$

and $N_{L/K}(\alpha) = a$. We then use the minimal equation for α over K. Let $r = \mathrm{ord}_K a$.

Case n/r. Choose $\pi \in K$ with ord $\pi = 1$. Then $a = \pi^r u$ where ord $u = 0$. If we find L and α with $N_{L/K}(\alpha) = u$, then $N_{L/K}(\pi^{r/n}\alpha) = a$. Thus it will suffice to do the case ord $a = 0$. Let L be the unramified extension of K of degree n (as in Lemma 9.22). By CL V § 2, $N_{L/K}: U_L \twoheadrightarrow U_K$ is onto so we can write $a = N_{L/K}(\alpha)$ with $\alpha \in L$. The problem is to choose α so $K(\alpha) = L$

If n is prime to the characteristic of K, there is again an easy proof. Let $L = K(\theta)$ for some θ. Since there are only a finite number of fields between K and L we can find arbitrarily large pairs $h < k$ with $K(\alpha + \pi^h\theta) = K(\alpha + \pi^k\theta)$. This field is then L since it contains θ. Choose h very large with $K(\alpha + \pi^h\theta) = L$. Then $b = N_{L/K}(\alpha + \pi^h\theta)$ is very close to a so we have $b^{-1}a = c^n$ for some $c \in K$. Therefore we can replace α by $c(\alpha + \pi^h\theta)$.

To avoid the assumption on n, we use a different method suggested by Eichler's proof. Let θ be a normal basis element as in Lemma 9.22 and let $\alpha_h = 1 + \pi^h\theta$ and $\xi_h = \alpha_h^{-1}\sigma(\alpha_h)$ where σ generates the galois group of L/K. We claim that $\xi_h^r\alpha$ will do for some h and r. Clearly $N(\xi_h\alpha) = N\alpha = a$. Fix h and consider the fields $K(\xi^r\alpha)$ where $\xi = \xi_h$. If m is the number of fields between K and L we will have $K(\xi^i\alpha) = K(\xi^j\alpha)$ for some $0 \leq i < j \leq m$. This field will contain ξ^{j-1} and so will contain

ζ^t where $t = m!$ We claim that for some h, $K(\zeta_h^t) = L$. It will then follow that $K(\zeta_h^i \alpha) = L$ for some i by the above argument.

Suppose the assertion is false so $K(\zeta_h^t) \neq L$ for all h. Thus, there will be some τ_h in the galois group G of L/K such that $\tau_h \neq 1$ and τ_h fixes ζ_h^t. Choose a subsequence of h such that $\tau_h = \tau$. Since τ fixes $\zeta_h^t = \alpha_h^{-t}\sigma(\alpha_h^t)$ we have $\alpha_h^t \tau\sigma(\alpha_h^t) = \sigma(\alpha_h^t)\tau(\alpha_h^t)$ as in Lemma 9.22. Now $\alpha_h^t = (1 + \pi^h\theta)^t$. Write $t = qs$ where q is a power of the characteristic of K and s is prime to q. Then $\alpha_h^t = (1 + \pi^{qh}\theta^q)^s$ so we have

$$(*) \quad (1+\pi^{qh}\theta^q)^s(1+\pi^{qh}\tau\sigma(\theta)^q)^s = (1+\pi^{qh}\sigma(\theta)^q)^s(1+\pi^{qh}\tau(\theta)^q)^s$$

Expand both sides by the binomial theorem, subtract 1, divide by $s\pi^{qh}$ and take the limit along the chosen subsequence of h. The result is $\theta^q + \tau\sigma(\theta)^q = \sigma(\theta)^q + \tau(\theta)^q$. Since q is a power of the characteristic, this implies $\theta + \tau\sigma(\theta) = \sigma(\theta) + \tau(\theta)$. As in Lemma 9.22, this is impossible unless char $K = 2$, $\sigma = \tau$, and $\tau\sigma = 1$. In this case the equation (*) becomes

$$(1 + \pi^{2qh}\theta^{2q})^s = (1 + \pi^{2qh}\sigma(\theta)^{2q})^s .$$

Therefore we see that $\eta_h = (1 + \pi^{2qh}\theta^{2q})^{-1}(1 + \pi^{2qh}\sigma(\theta)^{2q})$ is an s-th root of 1. There are thus at most s possible values for η_h in L. Since η_h gets arbitrarily close to 1 as $h \to \infty$ we see that $\eta_h = 1$ for large h (in the chosen subsequence). This implies that $\theta^{2q} = \sigma(\theta)^{2q}$ so $\sigma\theta = \theta$ since 2q is a power of the characteristic. This is a contradiction. Therefore our assertion is proved.

Case $n \nmid r$. Let $d = (n, r)$. Then $d \neq n$. Since d/r, the case already done shows that if E is the unramified extension of K of degree d, there is some $\beta \in E$ with $K(\beta) = E$ and $N_{E/K}(\beta) = a$. Since E/K is unramified, $r = \mathrm{ord}_K a = \mathrm{ord}_E a = d\ \mathrm{ord}_E \beta$ so $\mathrm{ord}_E \beta = r/d$. This is prime to n/d. Let $\pi \in K$ be as above, let α be a root of $x^{n/d} + \pi^h x + (-1)^{n/d}\beta$ and let $L = E(\alpha)$. The term $\pi^h x$ makes the equation separable. We choose h very large. Now $\alpha^{n/d} + \pi^h \alpha = \pm\beta$ in L. If we had $\mathrm{ord}_L(\alpha^{n/d}) \geqslant \mathrm{ord}_L(\pi^h \alpha)$ then $\mathrm{ord}_L \alpha \geqslant (n/d - 1)^{-1} \mathrm{ord}_L \pi \geqslant 0$. But then $\mathrm{ord}_L \beta \geqslant \min(\mathrm{ord}_L \alpha^{n/d}, \mathrm{ord}_L \pi^h \alpha) = h + \mathrm{ord}_L \alpha \geqslant h$ which is impossible for large h. Thus $\mathrm{ord}_L \alpha^{n/d} < \mathrm{ord}_L \pi^h \alpha$ and so $\mathrm{ord}_L \beta = \mathrm{ord}_L \alpha^{n/d}$ so $e \cdot r/d = n/d\ \mathrm{ord}_L \alpha$ where e is the ramification index of L over E. Since n/d is prime to r/d, we see that n/d divides e. But $ef = [L: E] \leq n/d$ so $[L: E] = n/d$ and $[L: K] = n$. Now $N_{L/E}\alpha = \beta$ so $N_{L/K}(\alpha) = a$. Finally, consider $K(\alpha)$. Since this contains $\beta = \pm(\alpha^{n/d} + \pi^h \alpha)$ we see that $K(\alpha) \supset E = K(\beta)$ so $K(\alpha) = E(\alpha) = L$.

If $a \in \mathfrak{o}$ then $f(x) \in R[x]$ automatically since f is irreducible. This is a well known consequence of Hensel's lemma W 2-2-3. We can give an alternative proof by noting that $f(x) = \prod (x - \alpha_i)$ where the α_i are the conjugates of α. Since the α_i are all conjugate under the galois group of the normal closure M of L over K, and there is only one extension of the valuation of K to M, we have ord α_i = ord α for all i. Thus ord a = n ord α so α and all α_i are integral if a is.

Appendix

For the reader's convenience, we collect here some of the basic algebraic results used in the notes. The elementary facts will be stated without proof. We then give a somewhat more detailed account of completion and the theory of Dedekind rings.

1. Localization. If S is a multiplicative subset of a commutative ring R (i.e., s, t $\in$ S $\Rightarrow$ st $\in$ S, and 1 $\in$ S) and M is an R-module we define M_S to be the quotient of $M \times S$ by the equivalence relation $(m, s) \sim (m', s')$ if $t(s'm - sm') = 0$ for some $t \in S$. Let $m/s \in M_S$ be the image of (m, s). With the usual rules for adding and multiplying fractions, R_S is a ring and M_S is an R_S-module. If Λ is an R-algebra and M is a Λ-module then Λ_S is an R_S-algebra and M_S is a Λ_S-module. Also $\Lambda_S \otimes_\Lambda M \xrightarrow{\approx} M_S$ by $\lambda/s \otimes m \mapsto \lambda m/s$. In particular, $R_S \otimes_R M \xrightarrow{\approx} M_S$. The functor $M \mapsto M_S$ is exact. Thus $R_S \otimes_R$ — is exact so R_S is a flat R-module. Similarly Λ_S is a flat Λ-module. Let $i: M \to M_S$ by $m \mapsto m/1$. Then $\ker i = \{m \in M | sm = 0 \text{ for some } s \in M\}$. If $X \subset M_S$ is an R_S (resp. Λ_S) submodule then $Y = i^{-1}(X)$ is an R (resp. Λ) submodule of M and $X = Y_S$. Thus every ideal J of R_S or Λ_S has the form I_S where I is an ideal of R or Λ. This gives a bijective correspondence between prime ideals of R_S and prime ideals of R not meeting S. If M is a right Λ module and N is a left Λ module, then $(M \otimes_\Lambda N)_S \approx M_S \otimes_{\Lambda_S} N_S$. If M and N are left Λ- modules and M is

finitely presented, i.e., $\exists \Lambda^m \to \Lambda^n \to M \to 0$, $m, n < \infty$ then $\mathrm{Hom}_\Lambda(M, N)_S \approx \mathrm{Hom}_{\Lambda_S}(M_S, N_S)$. This is a consequence of the following more general result.

Lemma A1. If R is a commutative ring, $R \to R'$ a homomorphism of commutative rings with R' flat over R, Λ is an R algebra, and M, N left Λ modules, then

$$R' \otimes_R \mathrm{Hom}_\Lambda(M, N) \longrightarrow \mathrm{Hom}_{R' \otimes_R \Lambda}(R' \otimes_R M, R' \otimes_R N)$$

is an isomorphism provided M is finitely presented over Λ.

This follows from the fact that it is true for $M = \Lambda$ using the left exactness of both sides as functors of M. The flatness of R' is needed to show the exactness of the left hand side.

If $\mathfrak{y}$ is a prime ideal in a commutative ring R, we write $M_{\mathfrak{y}}$ for $M_{R-\mathfrak{y}}$. If $M_{\mathfrak{m}} = 0$ for all maximal ideals $\mathfrak{m}$ then $M = 0$. It follows that a sequence of Λ-modules is exact if and only if its localization at each maximal ideal is exact. Since P is projective if and only if $\mathrm{Hom}_\Lambda(P, -)$ is an exact functor, a finitely presented Λ-module P is projective if and only if each $P_{\mathfrak{m}}$ is projective over $\Lambda_{\mathfrak{m}}$. If P is any projective Λ-module, P is a direct summand of a free module so P_S is projective over Λ_S.

Finally we note that if $R \subset R'$ are commutative rings, R_0 is the integral closure of R in R', and $S \subset R$, then $(R_0)_S$ is the integral closure of R_S in R'_S.

2. <u>Noetherian Rings and Modules</u>. If Λ is a ring, a Λ-module M is called noetherian if it satisfies the ascending chain condition on submodules. This is equivalent to the assertion that every non void collection of submodules has a maximal element or that every submodule is finitely generated. If $0 \to M' \to M \to M'' \to 0$ then M is noetherian if and only if M' and M" are. In particular, if Λ is left noetherian, i.e., noetherian as a left Λ-module, then every finitely generated left Λ-module is noetherian and is also finitely presented. If M is a noetherian Λ-module, then $M[t] = \Lambda[t] \otimes_{\Lambda} M$ is a noetherian $\Lambda[t]$-module. In particular, if Λ is left noetherian so is $\Lambda[t]$ (Hilbert's basis theorem). If R is a commutative ring and I is an ideal of R, it follows by Zorn's lemma that the set of prime ideals $\supset I$ has minimal elements. As a typical example of noetherian induction we show that if R is noetherian there are only a finite number of such primes minimal over I. If not, choose I maximal among ideals for which the theorem is false. We reach a contradiction by showing that I itself is prime. If $ab \in I$, $a, b \notin I$, let $I' = I + Ra$, $I'' = I + Rb$. Then $I' > I$, $I'' > I$ but any prime $P \supset I$ must contain I' or I".

Finally note that an increasing chain of submodules of M corresponds to a sequence of epimorphisms $M \to M_1 \to M_2 \to \cdots$. If M is noetherian such a sequence will eventually consist of isomorphisms. In particular, any epimorphism $M \to M$ will be an isomorphism.

3. Completion. Let Λ be a ring and J a 2-sided ideal of Λ. Then we define the completion of Λ at J to be $\hat{\Lambda} = \varprojlim \Lambda/J^n$. If M is a Λ-module, we define $\hat{M} = \varprojlim M/J^nM$. We say Λ (resp M) is complete at J if $\Lambda \to \hat{\Lambda}$ (resp $M \to \hat{M}$) is an isomorphism. Clearly $\hat{M}$ is a $\hat{\Lambda}$-module. In general, $\hat{M} \not\approx \hat{\Lambda} \otimes_{\Lambda} M$, $M \mapsto \hat{M}$ is not exact, and $\hat{\Lambda}$ is not flat over Λ. The following theorem enables us to establish these properties in certain cases.

Theorem A2 (Artin-Rees). Assume J is generated by a finite number of central elements of Λ. Let M be a noetherian left Λ-module and M' a submodule. Then there is a k with $J^nM \cap M' = J^{n-k}(J^kM \cap M')$ for $n \geqslant k$.

Proof. The Rees ring Γ of Λ is defined to be $\Gamma = \Gamma_0 \oplus \Gamma_1 \oplus \Gamma_2 \oplus \ldots$ where $\Gamma_0 = \Lambda_1$ $\Gamma_n = J^n$ and the multiplication $\Gamma_i \times \Gamma_j \to \Gamma_{i+j}$ is given by the ordinary multiplication $J^i \times J^j \to J^{i+j}$. Define a surjective map $\Lambda[t_1, \ldots, t_r] \to \Gamma$ by sending t_i to $j_i \in \Gamma_1 = J$ where $j_i \in J \cap \mathfrak{Z}(\Lambda)$ generate J. Since $\Gamma \otimes_{\Lambda} M$ is a quotient of $\Lambda[t_1, \ldots, t_r] \otimes_{\Lambda} M$ it is noetherian over $\Lambda[t_1, \ldots, t_r]$ and so over Γ. Since $\Gamma = \Lambda \oplus J \oplus J^2 \oplus \ldots$, we have $\Gamma \otimes_{\Lambda} M = M \oplus JM \oplus J^2M \oplus \ldots$. Let $N = M' \oplus (JM \cap M') \oplus (J^2M \cap M') \oplus \ldots$. This is a Γ-submodule of

$\Gamma \otimes_{\Lambda} M$ and so is finitely generated. Therefore it is generated by $M' \oplus (JM \cap M') \oplus \ldots \oplus (J^k M \cap M') = N_k$ say. Now if $n \geq k$, $J^n M \cap M' = \Gamma_n M \cap \Gamma N_k = \sum_{i=0}^{k} J^{n-i}(J^i M \cap M') = J^{n-k}(J^k M \cap M')$ since $J(J^i M \cap M') \subset J^{i+1} M \cap M'$.

Theorem A3. Assume J is generated by a finite number of central elements of Λ. If $0 \to M' \to M \to M'' \to 0$ is exact and M is noetherian then $0 \to \hat{M}' \to \hat{M} \to \hat{M}'' \to 0$ is exact.

If also Λ is left noetherian, then $\hat{\Lambda} \otimes_{\Lambda} M \to \hat{M}$ is an isomorphism for finitely generated M and $\hat{\Lambda}$ is flat as a right Λ-module.

Proof. The sequence $0 \to M'/M' \cap J^n M \to M/J^n M \to M''/J^n M'' \to 0$ is exact. Since $J^n M' \subset M' \cap J^n M \subset J^{n-k} M'$ by Theorem A2, we have $\varprojlim M'/M' \cap J^n M = \hat{M}'$. The lemma below now gives the first statement. The second follows as in Lemma A1. The first two parts show that $\hat{\Lambda} \otimes_{\Lambda} -$ preserves exactness for finitely generated modules. Any short exact sequence is a direct limit of finitely generated ones and $\hat{\Lambda} \otimes_{\Lambda} -$ preserves direct limits.

Lemma A4. Let $(0 \to A_n \to B_n \to C_n \to 0)$ be an inverse system of short exact sequences of modules indexed by the positive integers. If all the maps $A_{n+1} \to A_n$ are onto, then

$$0 \to \varprojlim A_n \to \varprojlim B_n \to \varprojlim C_n \to 0$$

is exact.

This is standard and easy.

By applying this lemma to $0 \to J/J^n \to \Lambda/J^n \to \Lambda/J \to 0$ we see that, without any hypothesis, $0 \to \hat{J} \to \hat{\Lambda} \to \Lambda/J \to 0$. More generally, $0 \to \widehat{J^n} \to \hat{\Lambda} \to \Lambda/J^n \to 0$ and

$0 \to \widehat{J^nM} \to \hat{M} \to M/J^nM \to 0$. However it is not clear that $\widehat{J^n} = \hat{J}^n$ or $\widehat{J^nM} = \hat{J}^n\hat{M}$.

<u>Corollary A5</u>. If Λ is left noetherian and satisfies the hypothesis of Theorem A3, and M is finitely generated, then $\widehat{J^nM} = \hat{J}^n\hat{M} = J^n\hat{M}$ and $\hat{M}/\hat{J}^n\hat{M} \approx M/J^nM$.

<u>Proof</u>. $M/JM = \hat{\Lambda}/\hat{J} \otimes_\Lambda M = \hat{\Lambda}/\hat{J} \otimes_{\hat{\Lambda}} \hat{\Lambda} \otimes_\Lambda M = \hat{\Lambda}/\hat{J} \otimes_{\hat{\Lambda}} \hat{M} = \hat{M}/\hat{J}\hat{M}$ but $M/JM = \hat{M}/\widehat{JM}$ so $\widehat{JM} = \hat{J}\hat{M}$. Use induction on n. Also applying $\hat{\Lambda} \otimes_\Lambda -$ to $0 \to J \to \Lambda \to \Lambda/J \to 0$ gives $0 \to \hat{\Lambda} J \to \hat{\Lambda} \to \Lambda/J \to 0$ so $\hat{J} = \hat{\Lambda} J$ and so is finitely generated.

<u>Corollary A6</u>. If Λ satisfies the hypothesis of Theorem A3, and if $0 \to M' \to M \to M'' \to 0$ with M noetherian, then M is complete if and only if M' and M" are. Therefore if Λ is also left noetherian and complete, then so are all finitely generated left Λ-modules.

This is immediate from the 5-lemma.

We say a right Λ-module M is faithfully flat if a sequence is exact if and only if its image under $M \otimes_\Lambda -$ is exact. This is so if and only if M is flat and $M \otimes_{\Lambda} N = 0$ implies $N = 0$. It is sufficient to consider cyclic modules N.

Corollary A7. If Λ is left noetherian and satisfies the hypothesis of Theorem A3, then $\hat{\Lambda}$ is faithfully flat as a right Λ-module if and only if J is contained in the Jacobson radical of Λ. In particular, this is so if Λ is complete.

Proof. If N is finitely generated and $\hat{\Lambda} \otimes_{\Lambda} N = 0$ then $N/JN = \hat{N}/J\hat{N} = 0$. If $J \subset \text{Jac.rad}(\Lambda)$ then Nakayama's lemma implies $N = 0$. If J is not contained in $\text{Jac.rad}(\Lambda)$, there is a maximal left ideal $M \not\supset J$. Let $S = \Lambda/M$. Then $JS = S$ so $S/J^nS = 0$ and $\hat{\Lambda} \otimes_{\Lambda} S = \hat{S} = 0$ while $S \neq 0$.

The hypothesis of Theorem A3 will hold if Λ is an R algebra (R a commutative ring), $I \subset R$ is a finitely generated ideal, and $J = \Lambda I$. We can always reduce to this case by taking $R = \mathfrak{Z}(\Lambda)$. If M is a Λ-module, then $J^nM = I^nM$ so the completions of M with respect to I and J are the same. In particular, if Λ is a complete R-module, then Λ is complete and so $J \subset \text{Jac.rad.}(\Lambda)$.

4. Dedekind Rings. A Dedekind ring is a commutative integral domain in which all ideals are projective. The obvious example is a principal ideal domain (PID) for which all ideals are free. If R is Dedekind so is R_S. A PID with only one non zero prime ideal is called a discrete valuation ring DVR. We will show that a local Dedekind ring is a DVR. If R is a DVR, let (p) be the unique

maximal ideal. The associated valuation is defined by $\text{ord}(r) = \sup\{n \mid p^n \mid r\}$. This satisfies $\text{ord}(xy) = \text{ord}\, x + \text{ord}\, y$, $\text{ord}(x + y) \geqslant \min(\text{ord}\, x, \text{ord}\, y)$, and $\text{ord}\, x = \infty \Longleftrightarrow x = 0$. To see that last, note R is noetherian (as a PID) and so $(x) < (p^{-1}x) < (p^{-2}x) < \ldots$ must stop. We extend ord to the quotient field K of R by $\text{ord}(x/y) = \text{ord}\, x - \text{ord}\, y$. The above properties still hold.

Definition. If R is an integral domain with quotient field K, a fractional ideal of R is an R-submodule I of K with $I \neq 0$ and $sI \subset R$ for some $s \neq 0$ in K. Define $I^{-1} = \{x \in K \mid xI \subset R\}$. Then I^{-1} is a fractional ideal and $II^{-1} \subset R$. We say I is invertible if $II^{-1} = R$. If I is finitely generated then $(I_S)^{-1} = (I^{-1})_S$ for any $S \subset R$.

The fractional ideals form a monoid under $(I, S) \longrightarrow IJ$ and the invertible ones form a group.

The following theorem compares the various possible definitions of Dedekind rings.

Theorem A8. Let R be a commutative integral domain. The following conditions are equivalent.

(1) R is a Dedekind ring.

(2) Every non zero ideal of R is invertible, i.e., the fractional ideals form a group.

(3) R is noetherian, integrally closed, and every non-zero prime ideal is maximal.

(4) R is noetherian and R_y is a DVR for all maximal ideals y.

(5) R_y is a DVR for all maximal ideals y and if $a \neq 0$, $a \in R$, there are only a finite number of maximal ideals containing a.

(6) Every ideal is a product of prime ideals.

(7) Every non-zero ideal is uniquely a product of prime ideals (up to order).

Proof. If I is invertible, $R = II^{-1}$ so $1 = \sum_1^n a_i b_i$ with $a_i \in I$, $b_i \in I^{-1}$. Let $f: I \to \coprod_1^n R$ by $f(x) = (xb_i)$ and $g: \coprod_1^n R \to I$ by $g((x_i)) = \sum x_i a_i$. Then $gf = id$ so I is projective and finitely generated. Conversely, if we have such f, g (even with infinite n) then g has the given form and so does f since a map $I \to R$ has the form $x \mapsto xb$ with $b \in K$. Since $f(1) = (b_i)$ only a finite number of $b_i \neq 0$. But $a_i \in I$, $b_i \in I^{-1}$, $\sum a_i b_i = 1$ so I is invertible. Thus (1) $\Leftrightarrow$ (2) (1) $\Rightarrow$ (4). We have just seen that R is noetherian. Since R_y is local, all ideals are free. There can be only one generator since for more generators a, b, ... we have ba - ab = 0, a non trivial relation. Thus R_y is a PID. (4) $\Rightarrow$ (1). Clearly $I_{\mathfrak{M}}$ is projective for all $\mathfrak{M}$. (4) $\Rightarrow$ (3). The last two conditions can be verified locally. They clearly hold for a DVR.

<u>(4) $\Rightarrow$ (5)</u>. Since all primes are maximal, the $y \supset Ra$ are minimal over it.

(5) $\Rightarrow$ (4). Let $I_1 \subset I_2 \subset \ldots$. We can assume $I_1 \neq 0$. Since R_y is noetherian we can find n so $I_{ny} = I_{n+1y} = \ldots$ for the finite number of y with $I_1 \subset y$. For the other y, $I_{1y} = \ldots = R_y$. Therefore $I_n = I_{n+1} = \ldots$.

(4) $\Rightarrow$ (7). If $I \neq 0$, define $\text{ord}_y I = \text{ord}_y a$ where $I_y = R_y a$. If $\text{ord}_y I \neq 0$ then $I \subset y$. Let $J = \prod y^{\text{ord}_y I}$, a finite product. Then $J = I$ since they are locally equal. If $I = \prod y_i^{n_i}$ then $n_i = \text{ord}_{y_i} I$ is unique. Clearly (7) $\Rightarrow$ (6). The fact that (6) $\Rightarrow$ (2) does not seem to come up in practice, the other conditions being easier to verify than (6). A proof will be found in Zariski-Samuel, Commutative Algebra, volume I.

It remains to show that (3) $\Rightarrow$ (4), i.e., that a local ring satisfying (3) is a DVR. Let $\mathfrak{a} = (a): b = \{r \in R \mid rb \in (a)\}$ be maximal among proper ideals of the form (a): b. If $xz \in \mathfrak{a}$ but but $z \notin \mathfrak{a}$, then $(a): xb \supset \mathfrak{a} + R_y > \mathfrak{a}$ so $(a): xb = R$. Thus $xb \in (a)$ so $x \in \mathfrak{a}$. This shows that $\mathfrak{a}$ is prime. By (3) the only primes are 0 and y, the unique maximal ideal so $y = \mathfrak{a} = (a): b$. Now $x = b/a \notin R$ otherwise $(a): b = R$. But $xy \subset R$ so $xy = R$ or $xy \lneq y$. In the second case x would be integral over R since y is a finitely generated R-module (let $y = (a_1, \ldots, a_n)$, $xa_i = \sum c_{ij} a_j$. Then $|x\delta_{ij} - c_{ij}| = 0$). Since R is integrally closed, $x \in R$, a contradiction. Thus $xy = R$ and $y = (p)$ is principal with

$p = 1/x$. If $\mathfrak{a}$ is any ideal, the chain $\mathfrak{a} < p^{-1}\mathfrak{a} < p^{-2}\mathfrak{a} < \ldots \subset R$ must stop, say at $p^{-n}\mathfrak{a}$. If $p^{-n}\mathfrak{a} \neq R$ it lies in (p) and $p^{-n-1}\mathfrak{a} \subset R$. Therefore $p^{-n}\mathfrak{a} = R$ and $\mathfrak{a} = (p^n)$.

The next theorem gives many examples of Dedekind rings.

Theorem A9. Let R be a Dedekind ring with quotient field K. Let L be a finite field extension of K. Let R' be the integral closure of R in L. Then R' is a Dedekind ring and KR' = L. If L/K is separable, R' will be a finitely generated R-module.

Proof. Since $R_S = K$, $S = R - \{0\}$ we have $KR' = R'_S = L$, the integral closure of K in L. By taking the integral closure in two steps it will suffice to do the separable case and the purely inseparable case. Since R' is integrally closed, we consider Theorem A8 (3). Let P' be a prime of R' and $P = R \cap P'$. Then R'/P' is integral over R/P. If $P \neq 0$, R/P is a field hence so is R'/P' and P' is maximal. We show $P \neq 0$ if $P' \neq 0$. Let $x \in P'$, $x \neq 0$. Let $x^n + a_1x^{n-1} + \ldots + a_n = 0$ where the $a_i \in R$ and n is minimal. Then $a_n \neq 0$. Clearly $a_n \in R \cap P'$.

We now assume L/K is separable and prove the last statement, which shows also that R' is noetherian. Let $w_1, \ldots, w_n$ be a base for L over K with $w_i \in R'$ (Note KR' = L since integral closure localizes). The map $L \to \prod_1^n K$ by $x \mapsto (tr(w_i x))$ is a monomorphism and hence an isomorphism since $\mathrm{tr}(w_i x) = 0$ all $i \Rightarrow \mathrm{tr}(yx) = 0$ all $y \Rightarrow \mathrm{tr}(z) = 0$ all $z \in L$ (a field) $\Rightarrow L/K$ is not separable. Choose $w_i' \in L$ so $\mathrm{tr}(w_i w_j') = \delta_{ij}$. If $x \in L$, $x = \sum_1^n c_i w_i'$ where $c_i = \mathrm{tr}(x w_i)$. If $u \in R'$, all conjugates of u are integral over R. Hence so is $\mathrm{tr}(u) \in K$ so $\mathrm{tr}(u) \in R$. Thus $x \in R'$ implies $c_i \in R$ so $R' \subset \sum_1^n R w_i'$, a finitely generated R-module.

We now consider the purely inseparable case. We use Theorem A8 (5). If char $K = p$, there is a $q = p^n$ so $x \in L \Rightarrow x^q \in K$. Suppose first that R is a DVR. Consider $L^* \xrightarrow{q \text{ power}} K^* \xrightarrow{\text{ord}} \mathbb{Z}$. We have $R' \cap K = R$ so $x \in R' \Leftrightarrow x^q \in R \Leftrightarrow \mathrm{ord}\, x^q \geq 0$. Let $x \in R'$ have least $\mathrm{ord}\, x^q > 0$. If $y \in R'$, then $\mathrm{ord}\, y^q = m\, \mathrm{ord}\, x^q$ some m so y/x^m is a unit of R'. From this it is trivial that R' is a DVR. In general, if P' is a non-zero prime of R', then $P = R \cap P' \neq 0$ (if $a \in P'$, $a^q \in P$). Since $R_P = R_S$ with $S = R - P$ is a DVR, so is R_S'. Thus so is $R_{P'}'$, a localization of R_S'. In fact $R_{P'}' = R_S'$. Finally let $a \neq 0$ in R'. If $a \in P'$, then $a^q \in P = R \cap P'$. There are only a finite number of possible P but P determines $P' = P_{P'}' \cap R' = R' \cap$ max.ideal of R_S'.

We now make a few general remarks on relatively prime ideals. Here R can be any ring, not necessarily commutative. We consider only 2-sided ideals. We say I and J are relatively prime if $I + J = R$. One useful consequence of this is that

$I \cap J = IJ + JI$. Clearly the right side always lies in $I \cap J$. If $I + J = R$, let $i + j = 1$. If $x \in I \cap J$, then $x = xi + xj \in JI + IJ$. If R is commutative, this reduces to $I \cap J = IJ = JI$. If I, J and I, J' are relatively prime, so are I, JJ' since the image of JJ' in R/I is the product of the images of J and J' both of which are R/I. In particular, I^m, J^n are relatively prime for all n, m. Note that I and J are relatively prime if and only if no maximal (2-sided) ideal contains both I and J.

<u>Theorem A10</u> (Chinese Remainder Theorem). Let M be any R-module. Let $I_1, \ldots, I_n$ be pairwise relatively prime. Then $M \to \prod M/I_iM$ is onto.

In particular, $R/I_1 \cap \ldots \cap I_n = \prod R/I_i$.

<u>Proof</u>. Fix i. For each j let $x_j + y_j = 1$ with $x_j \in I_i$, $y_j \in I_j$. Then $y_j \equiv 1 \bmod I_i$, $y_j \equiv 0 \bmod I_j$. Let $a_i = \prod_{j \neq i} y_j$. Then $a_i \equiv 1 \bmod I_i$, $a_i \equiv 0 \bmod I_j$ for $j \neq i$. If $m_i \in M$ let $m = \sum a_i m_i$. Then $m \equiv m_i \bmod I_i$.

<u>Corollary A11</u>. If I and J are relatively prime, then $I/I \cap J \approx R/J$.

<u>Proof</u>. $R/I \cap J \xrightarrow{\approx} R/I \times R/J$. The inverse image of $0 \times R/J$ is clearly $I/I \cap J$.

We now return to Dedekind rings. If $\mathfrak{a}$ and $\mathfrak{b}$ are fractional ideals of R write $\mathfrak{a} \sim \mathfrak{b}$ and say $\mathfrak{a}$ and $\mathfrak{b}$ have the same class if there is an $x \in K$ with $\mathfrak{b} = x\mathfrak{a}$. Then $\mathfrak{a} \sim \mathfrak{b}$ if and only if $\mathfrak{a} \approx \mathfrak{b}$ as R modules. In fact $f: \mathfrak{a} \to \mathfrak{b}$, applying $K \otimes -$ shows that $f(a) = xa$ for some $x \in K$. Note that $\mathfrak{a} \supset \mathfrak{b}$ if and only if $\mathfrak{a} = \mathfrak{b}\mathfrak{c}$ (with $\mathfrak{c} = \mathfrak{b}^{-1}\mathfrak{a} \subset R$). Thus two ideals are relatively prime if and only if their prime factorizations have no common y.

Theorem A12. Let R be a Dedekind ring. Let $\mathfrak{a}$ and $\mathfrak{b} \subset R$ be ideals. Then there is $\mathfrak{c} \subset R$ such that $\mathfrak{a} \sim \mathfrak{c}$ and $\mathfrak{c}$ is prime to $\mathfrak{b}$.

Proof. For each $y \supset \mathfrak{b}$, $\mathfrak{a}_y^{-1} = (\mathfrak{a}^{-1})_y = R_y a_y$ for some $a_y \in \mathfrak{a}^{-1}$. By Theorem A10 we can find $a \in \mathfrak{a}^{-1}$ so $a \equiv a_y \bmod y\mathfrak{a}^{-1}$. Thus a generator $\mathfrak{a}_y^{-1}$ by Nakayama's lemma since it generates $\mathfrak{a}_y^{-1}/y_y\mathfrak{a}_y^{-1}$. Let $\mathfrak{c} = a\mathfrak{a} \subset R$. If $y \supset \mathfrak{b}$ then $\mathfrak{c}_y = a_y\mathfrak{a}_y = \mathfrak{a}_y^{-1}\mathfrak{a}_y = R_y$ so $y \not\supset \mathfrak{c}$.

Corollary A13. Let $\mathfrak{a}$ and $\mathfrak{b} \subset R$ be any non zero ideals of a Dedekind ring R. Then $\mathfrak{a}/\mathfrak{a}\mathfrak{b} \approx R/\mathfrak{b}$.

Proof. This depends only on the isomorphism class of $\mathfrak{a}$. By Theorem A12 we can assume $\mathfrak{a}$ and $\mathfrak{b}$ are relatively prime. Apply Corollary A11.

Corollary A14. If $a \in \mathfrak{a}$, $a \neq 0$. Then there is $b \in \mathfrak{a}$ with $\mathfrak{a} = (a, b)$.

Let b generate the cyclic module $\mathfrak{a}/a\mathfrak{a}$. Then $\mathfrak{a} = a\mathfrak{a} + Rb$.

If all $R/\mathfrak{a}$ are finite for $\mathfrak{a} \neq 0$ we can define $N\mathfrak{a} = |R/\mathfrak{a}|$. Considering $R \supset \mathfrak{a} \supset \mathfrak{a}\mathfrak{b}$ and using

Corollary A13 we see that $N(\mathfrak{a}\mathfrak{b}) = N(\mathfrak{a})N(\mathfrak{b})$.

If $\mathcal{T}_R$ is the category of finitely generated torsion modules over R, then $K_0(\mathcal{T}_R)$ is free abelian with base $\{[R/y]\}$. If I_R is the group of fractional ideals of R, I_R is free abelian with base $\{y\}$. Let $\varphi : I_R \approx K_0(\mathcal{T}_R)$ by $\varphi(y) = [R/y]$. Then Corollary A13 shows that $\varphi(\mathfrak{a}) = [R/\mathfrak{a}]$ for $\mathfrak{a} \subset R$.

<u>5. Modules over Dedekind rings</u>. If Λ is a left hereditary ring, i.e., all left ideals are projective, then every projective module is a direct sum of ideals and every submodule of a projective module is projective. The proof may be found in Cartan-Eilenberg. We consider here the finitely generated case.

<u>Theorem A15</u>. Let R be a Dedekind ring. Then every finitely generated torsion free R-module $M \neq 0$ is projective and $M \approx F \oplus \mathfrak{a}$ where F is free and $\mathfrak{a}$ is an ideal. The rank of M and class of $\mathfrak{a}$ are unique.

<u>Proof</u>. Let M be such a module. Then $K \otimes_R M = K^n = K \otimes_R F$ where $F = R^n$ is free. Replace M by sM, $s \neq 0$ so $M \subset F$. Let $p: F \to R$ be projection on the last coordinate of F. Let $\mathfrak{a} = p(M) \subset R$. Then $0 \to M' \to M \to \mathfrak{a} \to 0$. Since $\mathfrak{a}$ is projective, $M = M' \oplus \mathfrak{a}$. By induction on $\operatorname{rk} M = \dim_K K \otimes_R M$ we see that M is a direct sum of ideals and hence projective. To get $M = R^{n-1} \oplus \mathfrak{a}$ we apply the following lemma.

<u>Lemma A16</u>. If $\mathfrak{a}$ and $\mathfrak{b}$ are non-zero ideals of a Dedekind ring R then $\mathfrak{a} \oplus \mathfrak{b} \approx R \oplus \mathfrak{a}\mathfrak{b}$.

Proof. We can assume $\mathfrak{a}$ and $\mathfrak{b}$ are relatively prime so $\mathfrak{a} + \mathfrak{b} = R$ and $\mathfrak{a} \cap \mathfrak{b} = \mathfrak{a}\mathfrak{b}$. The sequence $0 \to \mathfrak{a} \cap \mathfrak{b} \to \mathfrak{a} \oplus \mathfrak{b} \to \mathfrak{a} + \mathfrak{b} \to 0$ splits.

Finally we show the uniqueness of the class of $\mathfrak{a}$. That of rk F is clear. If $\Lambda(M)$ is the exterior algebra on M, then $\Lambda(M \oplus N) = \Lambda(M) \otimes_R \Lambda(N)$. Since Λ of a free module is free, this shows that Λ of a projective module is projective (let $M \oplus N$ be free and note $\Lambda(M) \otimes_R R$ is a direct summand of $\Lambda(M) \otimes \Lambda(N)$). For $n \geq 2$, $K \otimes \Lambda_n(\mathfrak{a}) = \Lambda_n(K \otimes \mathfrak{a}) = 0$ so $\Lambda_n(\mathfrak{a}) = 0$. Thus $\Lambda(\mathfrak{a}) = R \oplus \mathfrak{a}$ and $\Lambda^n(M) \approx \mathfrak{a}$.

As an application, consider the situation of Theorem A9 with L/K separable. Then R' is torsion free of rank $n = [L: K]$. If y is a non-zero prime of R, Theorem A15 and Corollary A13 show that $R'/yR' \approx (R/y)^n$. Let $R'y = p_1^{e_1} \dots p_s^{e_s}$. By Theorem A10, $R'/R'y = \prod R'/p_i^{e_i}$. Also $R'/p_i^{e_i}$ has a composition series with e_i factors R'/p_i by Corollary A13. If we let $f_i = [R'/p_i: R/y]$, this shows that $\sum_1^s e_i f_i = n$.

We now determine $K_0(R)$ and $G_0(R)$. These are equal since R is regular. Let I_R be the group of fractional ideals of R and P_R the subgroup of principal ideals Rx, $x \in K^*$. The class group of R is, by definition, $cl(R) = I_R/P_R$. We also define Pic R to be the group of isomorphism classes of finitely generated rank 1 projective modules with multiplication $[P][Q] = [P \otimes_R Q]$.

Theorem A17. If R is Dedekind, $G_0(R) = K_0(R) = \mathbb{Z} \oplus cl(R)$ and $\text{Pic } R = cl(R)$.

Proof. If M is a finitely generated projective module of rank n, define $\det M = \Lambda^n(M)$. Using $\Lambda(M \oplus N) = \Lambda(M) \otimes \Lambda(N)$ we see that $\det(M \oplus N) = \det(M) \otimes \det(N)$ so we have a homomorphism $\det\colon K_0(R) \to \text{Pic } R$. Since every finitely generated projective rank 1 module P is isomorphic to an ideal ($P \subset K \otimes_R P \approx K$) we see that $\text{Pic}(R) = cl(R)$. The multiplications agree since $\mathfrak{a} \otimes_R \mathfrak{b} \to R$ is a monomorphism with image $\mathfrak{a}\mathfrak{b}$. (Apply $K \otimes -$ to show that the kernel is torsion and thus 0 since $\mathfrak{a} \otimes_R \mathfrak{b}$ is projective.) Now rk, det: $K_0(R) \to \mathbb{Z} \times cl(R)$. The map $\mathbb{Z} \oplus cl(R) \to K_0(R)$ by $(n, [a]) \mapsto n[R] + [\mathfrak{a}]$ is clearly an inverse by Theorem A15.

The following theorem of Steinitz and Chevalley classifies all finitely generated R-modules.

Theorem A18. Let R be a Dedekind ring. Let $M \subset N$ be finitely generated torsion free R-modules. Then we can write $N = N_1 \oplus \ldots \oplus N_m \oplus N_{m+1} \oplus \ldots \oplus N_n$, $M = M_1 \oplus \ldots \oplus M_m$ where the M_i and N_i are isomorphic to ideals of R, $M_i = \mathfrak{a}_i N_i$ where $\mathfrak{a}_i \subset R$ and $\mathfrak{a}_1 \subset \mathfrak{a}_2 \subset \ldots \subset \mathfrak{a}_m$, $\mathfrak{a}_1 \neq 0$.

Any finitely generated R-module A has the form $A = R^n \oplus \mathfrak{a} \oplus R/\mathfrak{a}_1 \oplus \ldots \oplus R/\mathfrak{a}_m$ (or $M = R/\mathfrak{a}_1 \oplus \ldots \oplus R/\mathfrak{a}_m$ if M is torsion) where $0 \neq \mathfrak{a}_1 \subset \mathfrak{a}_2 \subset \ldots \subset \mathfrak{a}_m$. The ideals

$\mathfrak{a}_1, \ldots, \mathfrak{a}_m$, the integer n, and the class of $\mathfrak{a}$ are uniquely determined by M.

Proof. If A is an R-module let T be the torsion submodule. If A is finitely generated, $A/T = P$ is projective by Theorem A15 so $A = P \oplus T$. This shows the uniqueness of n and the class of $\mathfrak{a}$ in the second part since $T = R/\mathfrak{a}_1 \oplus \ldots \oplus R/\mathfrak{a}_m$. We must now show the existence of such a decomposition when $A = T$ is torsion. Let $\mathfrak{a}$ be the annihilator of A. We will show that there is an epimorphism $\eta : A \twoheadrightarrow R/\mathfrak{a}$. Let $\mathfrak{a} = \prod y_i^{n_i}$. If we find epimorphisms $\eta_i : A \twoheadrightarrow R/y_i^{n_i}$, then $\eta = (\eta_i) : A \twoheadrightarrow \prod R/y_i^{n_i} = R/\mathfrak{a}$ is an epimorphism since it is locally so. Let $y = y_i$, $n = n_i$, $\mathfrak{a} = y^n \mathfrak{b}$ where $\mathfrak{b}$ is prime to y. Then $\mathfrak{b}$ annihilates $y^n A$. If $y^{n-1}(A/y^nA) = 0$ then $y^{n-1}A \subset y^nA$ so $\mathfrak{b}\, y^{n-1}A = 0$ contradicting the choice of $\mathfrak{a}$. Therefore the annihilator of $B = A/y^nA$ is y^n. It will suffice to find $B \twoheadrightarrow R/y^n \twoheadrightarrow 0$. Write $B = F/N$ where F is a free R-module. Then $KN = KF$. In KN take $M = y^{-n}N$. Since $y^nF \subset N$ we have $M \supset F \supset N = y^nM$. By Theorem A15 and Corollary A13, $M/N \approx \prod_1^r R/y^n$ where $r = \operatorname{rk} N$. Thus $B = F/N \subset M/N = \prod_1^r R/y^n$. Let $\theta_i : B \twoheadrightarrow R/y^n$ be the i-th coordinate projection. If θ_i is not onto then $\theta_i B \subset y/y^n$ the unique maximal ideal (= unique proper submodule) of R/y^n. If this is so for all i, then $B \subset \prod_1^r y/y^n$ and so $y^{n-1}B = 0$, a contradiction.

We now have an epimorphism $\eta : A \to R/\mathfrak{a}$. Since A is an $R/\mathfrak{a}$-module and $R/\mathfrak{a}$ is a free module over this ring, η splits. Thus $A = R/\mathfrak{a} \oplus A'$. If $\mathfrak{a}'$ is the annihilator of A' then $\mathfrak{a} \subset \mathfrak{a}'$ and $A' = R/\mathfrak{a}' \oplus A''$. Repeating, we get $A = R/\mathfrak{a}_1 \oplus R/\mathfrak{a}_2 \oplus \ldots \oplus R/\mathfrak{a}_n \oplus A^{(n+1)}$ with $\mathfrak{a}_1 \subset \mathfrak{a}_2 \subset \ldots$. Since A is noetherian the increasing sequence $\sum_1^n R/\mathfrak{a}_i$ must stop.

To prove the uniqueness we can use induction on n, observe that $\mathfrak{a}_1$ is the annihilator of A and that A' in $A = R/\mathfrak{a}_1 \oplus A'$ is unique by the Krull Schmidt theorem. Alternatively we can use the Fitting invariants of A or we can observe that we can calculate $\operatorname{ord}_y \mathfrak{a}_i$ from a knowledge of $\dim_{R/y} y^n A/y^{n+1}A$ for all n.

We now prove the first part. This gives a slightly different proof of the existence of a decomposition for a finitely generated module. We use induction on the rank of N. If $A = N/M$, let $T = \text{Torsion}(A)$, $P = A/T$. Then $T = N'/M$ for some N' and $N/N' = P$. Thus $N = N' \oplus N''$ where $N'' \approx P$. By Theorem A15, $N'' = N_{m+1} \oplus \ldots \oplus N_n$ while $M \subset N'$. Thus we are done by induction unless $N'' = 0$ so that A is torsion. In this case let $\mathfrak{a}$ be the annihilator of A. By what we have shown above, there is an epimorphism $A \to R/\mathfrak{a}$. Since N is projective by Theorem A15 there is a map $\varphi : N \to R$ making the diagram

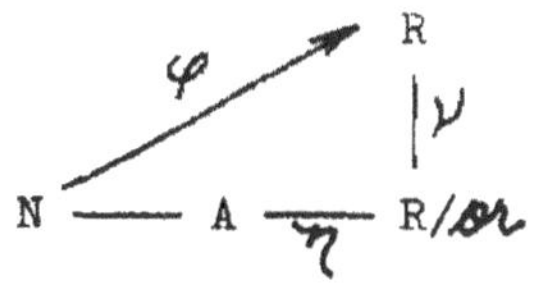

commute where $\nu(r) = r \bmod \mathfrak{b}$. Let $\mathfrak{b} = \varphi(N)$. Then $\nu(\mathfrak{b}) = R/\mathfrak{a}$ so $\mathfrak{b} \neq 0$. Let $\mathfrak{c} = \varphi(M) \subset \mathfrak{b}$. Since M has image 0 in A we have $\nu(\mathfrak{c}) = 0$. Since $\mathfrak{a}N \subset M$ we have $\mathfrak{a}\mathfrak{b} = \varphi(\mathfrak{a}N) \subset \varphi(M) = \mathfrak{c}$. If $\mathfrak{d} = \mathfrak{b} \cap \ker \nu$ then $\mathfrak{b}/\mathfrak{d} = \nu(\mathfrak{b}) = R/\mathfrak{a}$. But $\mathfrak{a}\mathfrak{b} \subset \mathfrak{c} \subset \mathfrak{d} \subset \mathfrak{b}$ and $\mathfrak{b}/\mathfrak{a}\mathfrak{b} \approx R/\mathfrak{a}$ by Corollary A13. If $\mathfrak{a}\mathfrak{b} \neq \mathfrak{d}$ we would get an epimorphism $R/\mathfrak{a} \to R/\mathfrak{a}$ with non-trivial kernel. This is impossible since $R/\mathfrak{a}$ is noetherian. Thus $\mathfrak{a}\mathfrak{b} = \mathfrak{c} = \mathfrak{d}$. Now $\mathfrak{b}$ is projective so we can find ψ: $\mathfrak{b} \to N$ with $\varphi\psi = \mathrm{id}_{\mathfrak{b}}$. Since $\varphi(M) = \mathfrak{c} = \mathfrak{a}\mathfrak{b}$, we have ψ: $(\mathfrak{c}) \subset \mathfrak{a}\psi(\mathfrak{b}) \subset \mathfrak{a}N \subset M$ so $\psi|\mathfrak{c}$ splits $M \to \mathfrak{c} \to 0$. Let $N_1 = \psi(\mathfrak{b})$, $M_1 = \psi(\mathfrak{c})$, $N' = \ker \varphi$, $M' = N' \cap M = \ker \varphi|M$. Then $N = N_1 \oplus N'$, $M = M_1 \oplus M'$, $M_1 \subset N_1$, $M' \subset N'$. Also $M_1 \subset N_1$ looks like $\mathfrak{c} = \mathfrak{a}\mathfrak{b} \subset \mathfrak{b}$ under φ so $M_1 = \mathfrak{a}N_1$. By induction we can put $M' \subset N'$ in the required form. Since $\mathfrak{a}N \subset M$ we have $\mathfrak{a}N' \subset M'$ so all ideals of $N' \subset M'$ will contain $\mathfrak{a}$.

As in the case of abelian groups, there is an alternative approach to the second part of Theorem A18 where we first decompose a torsion module into its primary component. We then regard the y-primary component as a module over R_y.

<u>Definition</u>. Let M be a torsion R-module. The y-primary component of M is $M^{(y)} = \{x \in M \mid y^n x = 0 \text{ for some } n\}$. We say M is y-primary if $M = M^{(y)}$.

<u>Lemma A19</u>. If M is y-primary then $M \to M_y$ is an isomorphism. If M is torsion and $M^{(y)} = 0$, then $M_y = 0$. Therefore if M is torsion then $M^{(y)} \xrightarrow{\approx} M_y$.

Proof. Let M be y-primary. If $x \in \ker[M \to M_y]$ let $\mathfrak{a}$ be the annihilator of x. Then $y^n \subset \mathfrak{a}$ and $sx = 0$ so $s \in \mathfrak{a}$ for some $s \notin y$. Thus $(y^n, s) \subset \mathfrak{a}$ but $(y^n, s) = R$ since no maximal ideal contains y^n and s. Suppose now M is torsion and $M^{(y)} = 0$. Let $x \in M$. Then $rx = 0$ for some $r \in R$. Let $(r) = \mathfrak{a} y^n$ where $\mathfrak{a} \not\subset y$. Then $y^n \mathfrak{a} x = 0$. Since $M^{(y)} = 0$, $\mathfrak{a} x = 0$. Since there is some $t \in \mathfrak{a}$, $t \notin y$, $tx = 0$ implies $x/s = 0$ in M_y. Finally, we localize $0 \to M^{(y)} \to M \to M/M^{(y)} \to 0$ at y and note that $(M/M^{(y)})^{(y)} = 0$. This gives $M^{(y)} \approx (M^{(y)})_y \approx M_y$.

Theorem A20. Let M be a torsion R-module. Then $M = \coprod_y M^{(y)}$.

Proof. The map $\coprod M^{(y)} \to M$ is locally an isomorphism by Lemma A19.

6. Global fields. A global field is either an algebraic number field (finite extension of $\mathbb{Q}$) or an extension of a finite field which is finitely generated and of transcendence degree 1. We want to classify all Dedekind rings whose quotient field is a global field.

Proposition A21. Let K be an algebraic number field and let R be the ring of integers of K, i.e., the integral closure of $\mathbb{Z}$ in K. If R' is a Dedekind ring with quotient field K, then $R' = R_S$ for some multiplicatively closed subset S of R.

Proof. Since R is finitely generated as a $\mathbb{Z}$-module (Theorem A9), the Jordan-Zassenhaus theorem applies and shows that the class group cl(R) is finite (only the very last part of the proof of the

Jordan-Zassenhaus theorem is needed. This is just the classical proof of the finiteness of class number). Now $\mathbb{Z} \subset R'$ and R' is integrally closed (Theorem A8) so $R \subset R'$. We now use the following lemma which makes no assumption on K and does not even require R' to be Dedekind.

Lemma A22. Let R be a Dedekind ring with quotient field K. Assume that cl(R) is a torsion group. If R' is any subring of K containing R, then $R' = R_S$ where $S = U(R') \cap R$.

Proof. Clearly $R_S \subset R'$ and $U(R') \cap R_S = U(R_S)$ since $r/s \in U(R')$, $r \in R$, $s \in S$ implies $r \in U(R')$ so $r \in S$. Replacing R by R_S we can assume that $U(R') \cap R = U(R)$. We must show that $R' = R$. Let $x \in R'$ and write $x = a/b$ where $a, b \in R$. Now $(a) = \prod y_i^{e_i}$, $(b) = \prod y_i^{f_i}$ in R so $Rx = \prod y_i^{e_i - f_i} = \mathfrak{a}\mathfrak{b}^{-1}$ say, where $\mathfrak{a}$ and $\mathfrak{b}$ lie in R and are relatively prime. Since cl(R) is torsion we can find an h with $\mathfrak{a}^h \sim R \sim \mathfrak{b}^h$. Let $\mathfrak{a}^h = (c)$, $\mathfrak{b}^h = (d)$. Those ideals are relatively prime since $\mathfrak{a}$ and $\mathfrak{b}$ are so we can write $gc + kd = 1$. Now $Rx^h = \mathfrak{a}^h \mathfrak{b}^{h^{-1}} = Rcd^{-1}$ so $x^h = ucd^{-1}$ where $u \in U(R)$. Therefore $u^{-1}gx^h + k = d^{-1} \in R'$. But $d \in R \subset R'$ so $d \in U(R') \cap R = U(R)$. Thus $\mathfrak{b}^h = (d) = R$ so $\mathfrak{b} = R$ and $Rx = \mathfrak{a}$. Therefore $x \in \mathfrak{a} \subset R$.

Remark. This is false without the hypothesis on cl(R). Let y be a prime ideal of R with $y^n \not\sim R$ for $n \neq 0$. Let $R' = \bigcup y^{-n}$. Then $U(R') \cap R = U(R)$. In fact if $x \in U(R') \cap R$ then $\mathrm{ord}_{\mathfrak{a}} x = 0$ for $\mathfrak{a} \neq y$ because $R'_{R-\mathfrak{a}} = R_{\mathfrak{a}}$. If $x \notin R$, then $\mathrm{ord}_y x = -n \neq 0$ so

$(x) = y^{-n}$ and $y^n \sim R$. For a treatment of the general case see W § 4-5.

We now turn to the case where K has characteristic $p \neq 0$.

<u>Proposition A23</u>. Let K be a global field over a finite field k. Let R' be a Dedekind ring with quotient field K. Then there is an $x \in R'$ such that $K/k(x)$ is a finite separable extension. If R is the integral closure of $k[x]$ in K then $R' = R_S$ where $S = U(R') \cap R$.

<u>Proof</u>. For the existence of x we only need to know that K is finitely generated over k, k is perfect, and R is any set generating K as a field over k. Since K is finitely generated, we can find a finite set $x_1, \ldots, x_n \in R$ such that $K = k(x_1, \ldots, x_n)$. Some $x = x_i$ must be transendental over k so $K/k(x)$ is algebraic. Choose such an x such that the largest possible number of x_j are separable over $k(x)$. Suppose $x_1, \ldots, x_r$ are separable over $k(x)$ and $x_{r+1}, \ldots, x_n$ are not. Let $y = x_{r+1}$ and let $h(Y) = 0$ be the minimal equation of y over $k(x)$. Clearing denominators we can assume that $h(Y) = f(x, Y)$ where $f(X, Y) \in k[X, Y]$ is irreducible. Since y is not separable over $k(x)$, we have $f(x, Y) = f_1(X, Y^p)$. If we had $f(X, Y) = f_2(X^p, Y^p)$ we could write $f(X, Y) = g(X, Y)^p$ using the fact that k is perfect to extract a p-th root of each coefficient of f. This would imply $g(x, y) = 0$, an equation of degree smaller than h. Thus f does not have the stated form. If $f(X, y)$ is irreducible over $k(y)$, this implies that x is separable over $k(g)$. But then so are $x_1, \ldots, x_r$ so we have a contradiction.

Suppose $f(X, y) = p(X, y)q(X, y)$. Since y is transcendental over k (otherwise it is separable over k (which is perfect)). Gauss' lemma shows that $f(X, Y)$ factors in $k[X, Y]$ which is again a contradiction.

The rest of the proof is just like that of Proposition A21. Combining Proposition A21 and A23 we get

Corollary A24. If R is a Dedekind ring whose quotient field is a global field then R satisfies the Jordan-Zassenhaus theorem.

We also remark that any integrally closed domain whose quotient field is a global field is a Dedekind ring. In fact the above proofs used only this property of R'.

Remark. If K is a global field over a finite field k, the field of constants k' of K is, by definition, the algebraic closure of k in K. This is also finite since a subfield of a finitely generated field extension is also finitely generated. We can also see this from Proposition A23 since $k[x] \subset k'[x] \subset R$ so $k'[x]$ is finitely generated as a $k[x]$-module.

We conclude by giving some alternative proofs of the strong approximation theorem (SAT) Theorem 9.21 . The following standard result shows a connection between the SAT and the theory of Dedekind rings.

Theorem A25. Let K be any field. Then there is a 1-1 correspondence between Dedekind rings R with quotient field K and sets E of valuations on K such that

(1) Every valuation $y \in E$ is discrete.

(2) If $x \in K$, $x \neq 0$, then $|x|_y = 1$ except for a finite number of y.

(3) E satisfies the SAT.

If R corresponds to E, then $R = \bigcap_{y \in E} \Theta_y$ where Θ_y is the valuation ring of y, and E is the set of valuations ord_y given by the DVR's R_y, $y \neq 0$.

Proof. If R is given, then $E = \{\mathrm{ord}_y\}$ clearly satisfies (1) and (2). (3) follows from the Chinese Remainder Theorem (CRT) if we observe that E satisfies the SAT if and only if given $y_1, y_2 \in E$, $y_1 \neq y_2$ and $E > 0$ we can find $\alpha \in K$ with $|\alpha|_{y_1} < \mathcal{E}$, $|1-\alpha|_{y_2} < \mathcal{E}$, and $|\alpha|_y \leq 1$ for the other $y \in E$. The proof is just like that of the CRT (Theorem A10). Clearly $E = \{\mathrm{ord}_y\}$ has this property by the CRT. By § 1, $R = \bigcap R_y$.

Given E, let $R = \bigcap_{y \in E} \Theta_y = \{x \in K \mid |x|_y \leq 1 \text{ all } y \in E\}$. If $x \in K$, but $x \notin R$, we find $s \in K$ with $|s-x^{-1}|_y < |x|_y^{-1}$ for $|x|_y > 1$ and $|s|_y \leq 1$ for the other $y \in E$. Then $s \neq 0$ and $sx \in R$ so K is the quotient field of R.. Let $P = R \cap y = \{x \in R \mid |x|_y < 1\}$. We claim $R_P = \Theta_y$. Clearly $R_{P_i} \subset \Theta_y$. If $x \in \Theta_y$ then $|x|_y \leq 1$. Choose s as above with $|s-1|_y < 1$. Then $sx \in R$ and $s \in R - P_i$. Now if $r \neq 0$, $r \in R$, then only a finite number of $R \cap y$ contain r by (2). If we show that every ideal lies in some $R \cap y$ then Theorem A8 (5) shows that R is Dedekind and we will have $E = \{\mathrm{ord}_P \mid P \text{ prime in } R\}$. Suppose $\mathfrak{A} \subset R$ but $\mathfrak{A} \not\subset R \cap y$ for all y. Then for each y, there is an $\alpha_y \in \mathfrak{A}$ with $|\alpha_y|_y = 1$. Using the SAT we can show $\mathfrak{A} = R$. See OM 22:1 for details.

Using this we will prove Theorem 9.21 for function fields. A function field K of dimension 1 over a field k is a finitely generated extension of k of transcendence degree 1. We do not require k to be finite here. The primes of K are by definition the valuations trivial on k. They are all discrete. Let Ω be the set of all primes of K. Then $k' = \bigcap_{y \in r} \Theta_y$ is the integral closure of k in K, i.e., the constant field.

<u>Theorem A26</u>. Let K be a 1-dimensional function field. Let E be any proper subset of Ω. Then E satisfies the hypothesis of Theorem A25.

Therefore E satisfies the SAT. Also $R = \bigcap_{y \in E} \Theta_y$ is a Dedekind ring with quotient field K.

<u>Remark</u>. This says in particular that if Γ is a nonsingular algebraic curve and $U \neq \Gamma$ is (Zariski) open in Γ then U is affine. The analogue in dimension 2 is false (remove $\Gamma \times P$ from $\Gamma \times \Gamma$).

<u>Proof</u>. Let $y_0 \in \Omega - E$. It will suffice to do the case $E = \Omega - \{y_0\}$. If n is large, dim ny_0 is large by the Riemann-Roch theorem (dim $\mathfrak{A}$ = deg $\mathfrak{A}$ + 1 - g + dim(k - $\mathfrak{A}$)). Thus there is a non-constant $x \in L(ny_0)$ i.e., $(x) + ny_0 \geq 0$. Clearly $x \in R = \bigcap_{y \neq y_0} \Theta_y$ so $k[x] \subset R$. Since R is integrally closed in K (as an intersection of DVR's), R contains the integral closure R' of k[x]

in K. By Theorem A9, R' is a Dedekind ring with quotient field K. Every prime in E comes from R' by Lemma A22 since $R' \subset R \subset \mathcal{O}_y$. Thus $E \subset \{\text{ord}_y \mid y \text{ prime in } R'\}$ which satisfies all the conditions by Theorem A25. (We conclude then that R' = R since the set of all primes Ω has $\bigcap_{y \in \Omega} \mathcal{O}_y = k'$.)

We now prove Theorem 9.21 for algebraic number fields. Let E consist of all primes but one y_0. Let T be a finite set of primes with $y_0 \notin T$. We are given $a_y \in K$ for $y \in T$ and must find $x \in K$ with $|x - a_y|_y < \varepsilon$ for $y \in T$ and $|x|_y \leq 1$ for $y \notin T$, $y \neq y_0$. By Theorem A25 we can find $y \in K$ with $|y - a_y|_y < \varepsilon/2$ for finite $y \in T$ and $|y|_y \leq 1$ for finite $y \notin T$. If $\mathfrak{A} \subset R$ is sufficiently divisible by all finite $y \in T$ and $z \in \mathfrak{A}$ then $x = y + z$ will satisfy $|x - a_y|_y < \varepsilon$ for finite $y \in T$ and $|x|_y \leq 1$ for finite $y \notin T$. If y_0 is finite, let $y_0^h = (s)$ and $S = \{s^n\}$. Then even $z \in \mathfrak{A}_S$ will do since we ignore y_0. We must pick z to make x behave properly at the infinite primes.

<u>Lemma A27</u>. If y is ∞, the image of $\mathfrak{A}$ in $\prod_{\substack{y \mid \infty \\ y \neq y_0}} \hat{K}_y$ is dense. If $S \not\subset U(R)$, the image of $\mathfrak{A}_S$ in $\prod_{y \mid \infty} \hat{K}_y$ is dense.

<u>Proof</u>. Let $A = \mathbb{R} \otimes_{\mathbb{Q}} K$. Then $\mathfrak{A}$ is a lattice in A so $A/\mathfrak{A}$ is compact. Now $A = \prod_{y \mid \infty} \hat{K}_y$. We want to show that $\mathfrak{A} + \hat{K}_{y_0}$ resp $\mathfrak{A}_S$ is dense in A. It will suffice to show that if X is a character of $A/\mathfrak{A}$ with $X(\hat{K}_{y_0}) = 1$ resp $X(\mathfrak{A}_S) = 1$, then $X = 1$. The trace

tr: $K \to \mathbb{Q}$ gives a non-degenerate bilinear form $t: K \times K \to \mathbb{Q}$ by $t(x, y) = tr(xy)$. This extends to $t: A \times A \to \mathbb{R}$ and $t(R \times R) \subset \mathbb{Z}$. This shows that $t': R \times A/R \to \mathbb{R}/\mathbb{Z}$ which gives a map $\Theta: R \to \widehat{A/R}$. If $\Theta(r) = 0$, then $t'(r, a) = 0$, i.e., $t(r, a) \in \mathbb{Z}$ for all $a \in A$ so $r = 0$ (t is $\mathbb{R}$-bilinear). Thus Θ is a monomorphism. But $R \approx \mathbb{Z}^n$, $A/R \approx (\mathbb{R}/\mathbb{Z})^n$ so $\widehat{A/R} \approx \mathbb{Z}^n$. Therefore $\Theta(R)$ has finite index in $\widehat{A/R}$ and so does $\Theta(\mathfrak{O})$ since $R/\mathfrak{O}$ is finite. Now $X \in \widehat{A/R}$ so $mX \in \Theta(\mathfrak{O})$ for some $m \neq 0$ in $\mathbb{Z}$. Since $\widehat{A/R}$ is torsion free, $mX \neq 0$ if $X \neq 0$. Let $mX = \Theta(c)$. Then $mX(x) = tr(xc)$ for all $x \in A$.

Identify A with $\prod_{y|\infty} \hat{K}_y$. Then $tr(x_{y_1}, \ldots, x_{y_r}) = \sum tr\, x_{y_i}$, c is identified with $(c, c, \ldots, c)$ and $\hat{K}_{y_0}$ with $\{(0, \ldots, 0, y)\}$. In the first case, we have $X(\hat{K}_{y_0}) = 0$ so $tr(0, \ldots, 0, cy) = tr(cy) \in \mathbb{Z}$ for all $y \in \hat{K}_{y_0}$. This implies $c = 0$ since $\hat{K}_{y_0} = \mathbb{R}$ or $\mathbb{C}$. In the second case, $X(as^{-n}) = 0$ for all $a \in \mathfrak{O}$, $n > 0$. Thus $tr(cas^{-n}) \in \mathbb{Z}$ for all $a \in \mathfrak{O}$. As in the proof of Theorem A9, $\{x \in K | tr(xa) \in \mathbb{Z}\} = \mathfrak{O}_0$ is a finitely generated R-module. All cs^{-n} lie in this. Let $\mathfrak{O}_0 = \sum_1^k Rw_i$. Then $x \in \mathfrak{O}_0$ implies $ord_{y_0} x \geq \min ord_{y_0} w_i$. But $ord_{y_0} cs^{-n} \to -\infty$ as $n \to 0$ unless $c = 0$.

References

[AT] E. Artin and J. Tate, Class field theory, Benjamin, New York.

[AG] M. Auslander and O. Goldman, Maximal orders, Trans. Amer. Math. Soc. 97 (1960), 1-24.

[BK] H. Bass, Algebraic K-theory, Benjamin, New York, 1968.

[C] C. Chevalley, L'arithmétique dans les algèbres de matrices, Act. Sci. Ind. 323, Hermann, Paris, 1936.

[CL] J-P. Serre, Corps locaux, Act. Sci. Ind. 1296, Hermann, Paris, 1962.

[CR] C. W. Curtis and I. Reiner, Representation theory of finite groups and associative algebras, Interscience, New York 1962.

[DA] M. Deuring, Algebren, Springer, Berlin, 1935.

[DR] A. Dress, On the decomposition of modules, Bull. Amer. Math. Soc. 75 (1969), 984-986.

[E] M. Eichler, Uber die Idealklassenzahl hyperkomplexer Zahlen, Math. Z. 43 (1938), 481-494.

[J] H. Jacobinski, Genera and decompositions of lattices over orders, Acta Math. 121 (1968), 1-29.

[L] T. Y. Lam, Induction theorems for Grothendieck groups and Whitehead groups of finite groups, Ann. Ec. Norm. Sup. (4) 1(1968), 91-148.

[OM] O. T. O'Meara, Introduction to quadratic forms, Springer, Berlin, 1963.

[R] A. V. Roiter, On integral representations belonging to a genus, Izv. Akad. Nauk. SSSR 30 (1966), 1315-1324.

[RS] I. Reiner, A Survey of integral representation theory, Bull. Amer. Math. Soc. 76 (1970), 159-227.

[SK] R. G. Swan, Algebraic K-theory, Lecture notes in math. 76, Springer, Berlin, 1968.

[SP] R. G. Swan, Projective modules over group rings and maximal orders, Ann. of Math. 76 (1962), 55-61.

[W] E. Weiss, Algebraic number theory, McGraw-Hill, New York, 1968.

[ZS] O. Zariski and P. Samuel, Commutative algebra, vol. I, van Nostrand, Princeton, 1958.

List of Symbols

Index